To Dad June 07
Happy Birthday
Hopefully this will answer all your questions
lots of love
Dawn + Cliff

THE FARMING
OF NEW ZEALAND

THE FARMING OF NEW ZEALAND

The people *and the land*

GORDON McLAUCHLAN

NEW PHOTOGRAPHY BY IAN BAKER

PENGUIN
VIKING

To my whanau: none of you are famous and that doesn't matter,

but you are all decent and useful citizens and that does.

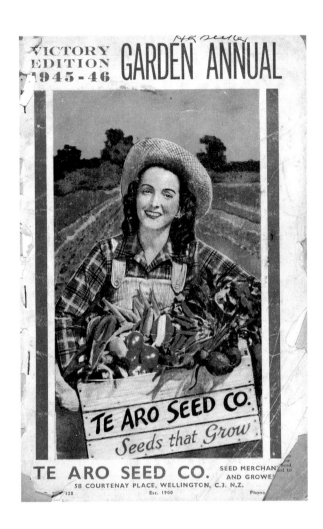

VIKING
Published by the Penguin Group
Penguin Group (NZ), 67 Apollo Drive, Mairangi Bay,
Auckland 0632, New Zealand (a division of Pearson New Zealand Ltd)
Penguin Group (USA) Inc., 375 Hudson Street,
New York, New York 10014, USA
Penguin Group (Canada), 90 Eglinton Avenue East, Suite 700, Toronto,
Ontario, M4P 2Y3, Canada (a division of Pearson Penguin Canada Inc.)
Penguin Books Ltd, 80 Strand, London, WC2R 0RL, England
Penguin Ireland, 25 St Stephen's Green,
Dublin 2, Ireland (a division of Penguin Books Ltd)
Penguin Group (Australia), 250 Camberwell Road, Camberwell,
Victoria 3124, Australia (a division of Pearson Australia Group Pty Ltd)
Penguin Books India Pvt Ltd, 11, Community Centre,
Panchsheel Park, New Delhi – 110 017, India
Penguin Books (South Africa) (Pty) Ltd, 24 Sturdee Avenue,
Rosebank, Johannesburg 2196, South Africa

Penguin Books Ltd, Registered Offices: 80 Strand, London, WC2R 0RL, England

First published by Penguin Group (NZ) in 2006
1 3 5 7 9 10 8 6 4 2

Copyright © text Gordon McLauchlan, 2006
Copyright © illustrations remains with the individual copyright holders

The right of Gordon McLauchlan to be identified as the author of this work in terms of section 96 of the Copyright Act 1994 is hereby asserted.

All rights reserved. Without limiting the rights under copyright reserved above, no part of this publication may be reproduced, stored in or introduced into a retrieval system, or transmitted, in any form or by any means (electronic, mechanical, photocopying, recording or otherwise), without the prior written permission of both the copyright owner and the above publisher of this book.

Designed by Dexter Fry
Printed by Everbest Printing Co. Ltd, China

ISBN-13 987 0 67 004562 4
ISBN 0 670 004562 4
A catalogue record for this book is available
from the National Library of New Zealand.

www.penguin.co.nz

CONTENTS

	Foreword	7
Introduction	Farming is For Ever	8

PART ONE

Chapter One	Pre-European Farming	14
Chapter Two	The New Crops, Implements and Animals	21
Chapter Three	A Lanky Land	32
Chapter Four	The First European Farmers	38
Chapter Five	Sheep in the Middle Island	50
Chapter Six	Grain and Ingenuity	64
Chapter Seven	Unwelcome Immigrants	78
Chapter Eight	Refrigerated Shipping	90
Chapter Nine	Each Day Was 'Just So Many Trees'	104
Chapter Ten	The Dairy Industry	116
Chapter Eleven	The Great Ideal – Land of One's Own	133
Chapter Twelve	The Marriage of Farming and Science	140

PART TWO

Chapter One	The Farmer's War, and the Aftermath	150
Chapter Two	Cast Off After Ninety Years	160
Chapter Three	Food and the Social Revolution	170
Chapter Four	Opening the World to New Zealand Milk	183
Chapter Five	The Re-emergence of Lamb	200
Chapter Six	The Vagaries of Beef	212
Chapter Seven	Deer and Goats	220
Chapter Eight	The Fall and Rise of Forests	228
Chapter Nine	New Fruit for Old	234
Chapter Ten	The Vegetable Business	248
Epilogue	The Future: A Guessing Game	254
	Acknowledgements	260
	Index	261

FOREWORD

As the market leader and largest rural bank in New Zealand, The National Bank is the proud sponsor of Gordon McLauchlan's *The Farming of New Zealand*.

This book chronicles the path of farming from pre-European times, following the development and innovation that has brought New Zealand to the enviable global position it enjoys today.

New Zealand's rural industry is fundamental to our economy and The National Bank has been a key driver in helping this industry grow. We work in partnership with our rural clients and provide the services, knowledge and specialist advice that helps build and enhance their businesses.

Supporting this book gives us an opportunity to both strengthen our relationship with, and give something back to, the rural community.

The Farming of New Zealand is a showcase of our vital farming industry, and has something for all New Zealanders. We hope you enjoy it.

Graham Hodges
Chief Executive and Director
ANZ National Bank Limited

INTRODUCTION

FARMING IS FOR EVER

Not long ago, newspapers carried a New Zealand Press Association story about a farmer who was 'rammed' against the 'rails of a pen' by a bull immediately after he had inserted an ear tag. According to a rescue helicopter spokesman, the farmer 'ended up with one of the horns between his legs which missed everything' – 'everything' being a euphemism I'm sure the farmer enjoyed once he recovered from 'the serious bruising to his lower back and spine'. But what made me laugh out loud was the last sentence of the story: 'It was unclear why the bull rammed the man.'

A bull is one kind of ruminant, the sort that chews its cud, but it is not the other, rational kind which the *Concise Oxford Dictionary* defines as 'given to meditation'; so I'm sure the cause of his attack was also unclear to the bull.

The journalist who wrote the story was clearly a city-dweller, and if you live in a city nowadays – especially a biggish one like Auckland – you see the grass in such small patches you forget that it's the stuff that cud is made of, the stuff that has largely supported the economy of this country for a good 160 years.

No one of my generation in any walk of life could possibly have escaped the knowledge that New Zealand earned its living in the world by selling wool, meat, dairy produce, and apples and pears. You didn't just learn it at school, it pressed on your life. The seas of grass lapped into the suburbs, straggling around sagging wire fences. We played football after school among the sheep's tollies and the cowpats. If cuts on arms and legs sometimes swelled with pus, stinging hot poultices drew the poison out. I feel a great unease now as some of the finest grassland in the world recedes, yielding to urban subdivisions and their barren, monotonous houses that will one day spawn barren, monotonous children.

From the ages of eleven through fourteen I spent summer holidays and many weekends milking cows and haymaking in shorts and sandals and not much else: red skin peeled and, after a week or two of each summer, left a heavy brown shell as well as unseen scars that blossom now as basal cell carcinoma. On the twice-daily train between the Hutt Valley and Wellington College the stink would punch into my nostrils as I travelled past the Ngauranga freezing works at the foot of the gorge, but I didn't recoil then as I would now in a more sanitised world. A few years later, for three summer seasons, I worked, enveloped in that pong but not noticing it any more, on the end of a butchers' chain, fit but too small for the chain itself. Powerful men sweated through eight-hour days at piece rates. The contract chains went only as fast as the slowest links, so the weak and the slack were soon prised out.

For overtime at night, we worked in teams of four folding lambs' legs through the brisket and pulling the shortened carcass into a muslin bag before it was frozen. 'Telescoping', it was called. A process designed during the war to save shipping space, it survived the war for a decade or more. They were stumpy lambs then, swathed in layers of white fat, not like the lean carcasses of today. We students worked with common men who often

Freezing workers in 1948 handling carcasses – 'telescoped' to save shipping space – as they came from the freezer ready for loading on refrigerated ships. We worked with common men, who often turned out to be not so common, folding lambs' legs through the brisket and pulling the shortened carcass into a muslin bag before it was frozen.

turned out to be not so common at all. I talked for several nights at the telescoping table with a forty-plus Englishman who went to sea at thirteen, spent his off-time reading American novels, and was one of the best-read men I've ever met – able to quote passages from Sherwood Anderson, Jack London, Sinclair Lewis, Ernest Hemingway, F. Scott Fitzgerald, Upton Sinclair, and more. He had jumped ship in New Zealand as so many did after the war and was lying low until the inevitable amnesty came. On the hottest days we would sometimes swim off the rocks at Ngauranga during the tea break, checking first which way the sea was sweeping the blood and guts from the effluent pipe, so we could swim up-tide and avoid the unpleasant viscera.

Full-time university students were a rarity, the sons and daughters of the extremely privileged, unenvied, just lazy to us, and they certainly missed out on the richest part of our education – as all students do today. Those students not in the freezing works or on the wharves over summer mostly worked in the woolstores.

City workers who idled away a sunny lunchtime with a walk on the wharves would see almost nothing but wool bales, carcasses and butter boxes being craned aboard the 'Home' boats. When I worked shifts on a news agency I seagulled on the wharves during the four-day breaks as the shifts changed, even after I was married, because the money was good. That's all gone now, and it went very quickly in the 1960s and 1970s. Town and country had not always lived easily together but their lives did then interlock.

For several years in the 1960s I spent a lot of time driving around New Zealand on assignment for the *New Zealand Weekly News*, most often called the 'Auckland Weekly', and for the *New Zealand Journal of Agriculture*, of which I was editor for four years. I gloried in the New Zealand countryside: from the backblocks of Canterbury and the rough, broken landscapes of Molesworth Station and Central Otago (worked by the hardest of men) to the kempt, almost manicured orchards and stud farms in places like the Wairarapa, Manawatu and Bay of Plenty. I also enjoyed farmers: usually taciturn, easy-mannered men with an earthy sense of humour; often slow to speak but firm to the point of obduracy in their opinions once expressed; always sizing up townies, looking for the mental and moral flab they were pretty sure they would find sooner or later.

I remember too the competence of their wives at what is now called multitasking but had no name

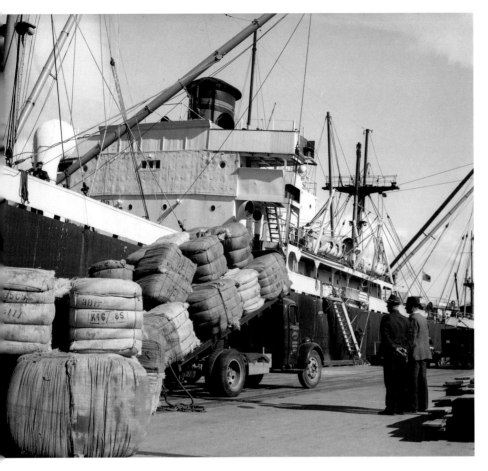

Bales of wool being loaded on an export ship in 1950 while Wellington city workers watch.

then. They just did all those things because there was no one else to do them. And many of the children fled the farm for the professions in town because it was such an unremittingly demanding life, but they took with them the confidence and maturity that hard work had lent them. I smile when I think of the gorgeous young woman home on holiday from university to the family Romney stud, emphasising to a visiting group of foreign farmers that the ram that had just serviced a ewe with such alacrity in the shed was typical. 'All Dad's rams jump like that,' she said proudly.

The history of New Zealand economic life hung in the air and was stamped on the land by dilapidated cowsheds, shearing sheds, barns, and old milk-stands at the gate, replaced but not torn down yet; by old reapers, traction engines, tractors, small trucks, partly cannibalised and waiting patiently for a decent grave in case their bones could still be picked. Farmhouse kitchens and laundries were being remodelled at last and wives were relishing the appliances, if not quite trusting them yet to properly clean and dry, so the coppers and lead-lined tubs were still in the old washhouse, 'in the meantime', which meant for years.

My two magazines had huge circulations by today's standards, bringing the country to town and vice versa. Long before business pages graced the daily newspapers, agricultural journalists were among the top practitioners, the ones who the front page was often held for because the impact of what happened on farms, in farm produce processing plants and in the markets hit home at the financial well-being of us all. It still does, but nowadays not only journalists have averted their eyes from the rural economy and our economic history. Since the 1960s, conclaves of economists, prolixities of politicians and seminars of suits have been exhorting New Zealanders to adopt an entrepreneurial spirit, to take risks for the good of the country and the glory of the Organisation for Economic Co-operation and Development; and hardly any of these people seem to have noticed that the great entrepreneurs of New Zealand history have been its farmers. Although they have sometimes been lulled into somnolence during New Zealand-Britain, government-to-government marketing arrangements, they have survived the strife of floods and frosts and snow that may kill their capital assets, and the unpredictability of commodity prices as hard to manage in only a season or two as turning a battleship around at full speed.

But their greatest achievement has been largely forgotten. In 1986 the government abruptly pulled all the financial subsidies that had helped keep farmers afloat and, previous governments had thought, kept the New Zealand economy on an even keel. It was trauma time in the country for several years from 1985, a time of protest marches, bankruptcies and even suicides. A visiting English agricultural journalist, Claire Powell, wrote to her home paper: 'Farmers are reluctantly reading the writing that has been on the wall for some time – that their traditional farm produce is in a state of world-wide oversupply; their home market of 3.1 million is minute and their God-given rights to vital overseas markets are weakening'.

She was right. Commercial lending rates were around twenty-five per cent and rising, and inflation at more than ten per cent pushed up costs while farmers' gross incomes were nosediving. The government didn't care much because it thought the heydays of pastoral farming were over, and knew voting power now resided in the towns. Just as farmers were told to expect a forty per cent income slide, public servants were awarded pay increases of thirty-seven per cent. Some highly efficient young farmers with traditionally high debt load at the beginning of their working lives were whacked into submission. Once more, Claire Powell wrote with the portentousness we all felt: 'The main victims of the present crisis are the enthusiastic, heavily mortgaged young men, often on as much as eighty-five per cent borrowed money. They are being the first to be forced from the land. There are few willing or able to fill their shoes under present circumstances. Only time will tell if a vacuum of competent farmers is being created.'

Well, time has told. In 1980 traditional agriculture earned seventy-one per cent of New Zealand's merchandise export income. With forestry and fishing added, the percentage was eighty-three. In 2005 – twenty years after many economists and politicians had declared traditional agriculture would not sustain the New Zealand economy in the twenty-first century – earnings still represented fifty-five per cent of merchandise export earnings and had held at that level for fifteen years. With forestry and fishing added, the figure was seventy per cent, again steady for fifteen years. Although tourism was burgeoning and some small industries were growing slowly, primary producers were still keeping the country a going concern.

The market spread demonstrated that primary producers and processors were matching sophisticated products with widespread international demand. In 2005 Australia was the biggest customer, with the United States and Japan not far behind. The United Kingdom's share had slumped to five per cent by 2000, down from sixty-six per cent just after the war. But, remarkably, a consistent forty-four per cent was sold to 'other countries', too numerous to mention.

During the fifteen years from 1990, farming productivity increased dramatically, spectacularly so in some sectors such as dairying. Farmers and their advisers continued to test with great energy prospects for products, for markets, and for more efficient ways of using their land and management resources. Horticulturists had taken large risks in search of market niches. Some had been burnt but most had experienced the sort of success that rewarded their daring.

A measure of the recent achievement of farming and the confidence its success had built in its future was the increase in the value of rural land: it doubled between 1990 and 2000, and doubled again between 2000 and 2005. This recovery from the mid-1980s – when investment in farming and its infrastructure almost stopped – had some economists asking if the land price was not dangerously high considering the return on capital in most agricultural enterprises. But there was no stemming the flow of investment into the primary industries, and until the economy becomes more balanced with an expansion of the manufacturing and service industries, food and fibre production and processing will continue to keep the country in business.

All primary production remains high-risk, with periodic overproduction and consequent price slumps a necessary accompaniment of the farming business. But with a rare combination of doggedness and innovation the men and women of the land have prevailed once again and left many of us confident that, for New Zealand anyway, farming is for ever.

Minister of Finance Roger Douglas reads the 1985 Budget which included a packet of reforms to deregulate the economy and heralded trauma time for farmers.

A dilapidated barn near Waimate North: the history of New Zealand economic life hung in the air and was stamped on the land by old cowsheds, shearing sheds, barns and old milkstands, replaced but not yet torn down . . .

PART ONE

CHAPTER ONE

PRE-EUROPEAN FARMING

The cabbage tree required a lot of labour to dig out, cut and cook its elongated root, and was eaten as a delicacy rather than a staple food.

The original Maori kumara had a slim white tuber, although it was very nutritious. However, both the introduced red kumara and especially the potato were more prolific. The red kumara was brought in by Europeans in the first few years of the nineteenth century and quite soon produced outstanding harvests.

POLYNESIANS WERE the nomads of the Pacific and for thousands of years they had it to themselves, always on the move, but pausing here and there, maybe for decades, maybe for centuries. The reasons for their journeys of exploration were complex but included some deep cultural longing to move on as well as the smallness of their island homelands which constricted population growth and exacerbated political and religious tensions. The sea, after all, was as much their home as the land. So they were profoundly experienced and adept at the business of packing and leaving, and then arriving somewhere new.

All the way across the Pacific, from the west to the east, the Polynesians took with them and successfully transplanted their tropical plants – until they came to temperate zone New Zealand. Of the other permanently settled Polynesian Islands, only Rapa Nui (Easter Island) lies below the Tropic of Capricorn and it is not as far south as New Zealand. On some tropical atolls the soil would not support some of the richer food plants – breadfruit, for example – although the ubiquitous coconut seldom, if ever, failed them. So while the first Polynesian settlers came to New Zealand from the Arcadian islands of the tropical Pacific into an alien environment with plants that failed to survive at worst, and barely clung to life at best, they were quickly and comprehensively dispossessed a few centuries later by Europeans who found here a land and climate ideally suited to the crops and animals they brought from their own, more distant Hawaiki.

The early migrants to New Zealand must have been dismayed to find the colder temperatures would not allow them to enjoy the prodigality of the coconut, the banana or the breadfruit. Of the other plants they brought with them, the kumara would flourish only in northern coastal regions, but would grow most years as far south as Banks Peninsula if carefully tended. Taro grew only in some parts of the north, needed much attention and even then took a long time to mature. The gourd (hue) was propagated from seed (the only Maori staple that was) and eaten as a young plant in the period before the kumara was ready. Many of them were also allowed to mature and were then hardened in the sun or by fires and used as vessels to hold water, oils and other liquids, as well as birds and fish preserved in fat. Another use was as floats. The gourd's flesh was thin, like the European marrow, and of less nutritional value than the kumara, taro or yam. The yam (uhi or uwhi) was grown extensively in the Bay of Plenty and was common on the East Coast and other warm coastal regions of the north and needed much attention, but did not thrive elsewhere. The yam languished and then disappeared after the introduction of the potato.

The tropical cabbage tree (ti pore), also believed to be brought in by early Polynesian immigrants, was planted by Maori near settlements but never became prolific. The most prized of a number of cabbage tree varieties growing here, ti pore required a lot of labour to dig out, cut and cook its large elongated root, and was eaten as a delicacy rather

than a staple food. Karaka trees were planted by Maori near their settlements for their fruit, but not cultivated in plantations, just as flax (*Phormium tenax*) was planted within easy range of the community. The paper mulberry (aute) did not grow well and evidence suggests that attempts were first made to substitute the felted clothing material prepared from the aute with the bark of the 'lacebark' (*Hoheria*) before the wonders of flax were discovered.

The livestock most expeditions would have set out with included pigs, fowl, dogs (kuri) and rats (kiore). Pigs and fowl never made it all the way – or at least fowl may have but were less valued as food by the newcomers than the meaty and easily hunted land and sea birds that flocked in huge numbers throughout the country. Dogs were the only animal they domesticated. Thus, the earliest arrivals from Polynesia became hunters and gatherers rather than the gardeners they had been on their previous home islands. Moa, kiwi and other plump land birds unused to predators were easy pickings and became rare within a relatively short time, not only because of overhunting but also because of environmental changes, such as deforestation, wrought by these first humans to arrive. Seals and a variety of fish, shellfish and sea birds also became everyday foods. Basic carbohydrate was the fern root, aruhe (the rhizome of *Pteris aquilina*), which was available in most parts of the country. It was dug up and ground into a rough powder, and easily stored.

Although the forest was a source of timber for canoes, buildings, weapons, and some food – edible berries, young palm fronds, the bulbous upper stem of the nikau palm, and birds – it could not and did not sustain dense populations. Therefore, destruction of large areas by burning was not lamented by

In early spring teams of men attacked the area to be planted with the ko, or digging stick, and worked rhythmically together.

Maori intent on encouraging new fern growth and the development of kumara gardens as the fertility of old ones was expended. Because of the difficulty of controlling burn-offs, many fires would have spread far beyond the intentions of Maori gardeners. One estimate is that about half the area covered by forest when the Polynesians arrived had been destroyed by the time Pakeha came to settle. Some forests, especially in low-rainfall regions on the East Coast, never regenerated beyond native grasses and scrub. It was on this land that the first Pakeha farmers ran their sheep. Burning and tree-felling with iron tools continued after the prolific potato was introduced by Europeans who, intent on pastoral farming, became even more efficient at denuding the land of forest than their predecessors. Well before the end of the nineteenth century reafforestation had become a political issue.

The profligacy of early Maori with resources is not hard to understand. They could not garden the munificent plants of the tropics, and land birds and bracken were readily available in a country that seemed huge, and was, in fact, bigger than all the other islands of Polynesia put together. The meaty moa and seals were easy to catch and kill, and for a long time must have seemed an infinite resource until they became, in some cases, close to extinction. Hence, as the land birds dwindled and seals retreated to more isolated beaches, cultivation increased in importance.

By the time James Cook arrived, the shortage of land birds meant Maori – an estimated 100,000 to 150,000 of them – were farmers again, with the main centres of population gathered around the cultivation of the kumara in the coastal regions of the North Island, or inland around the thermal areas of the Bay of Plenty and along the fertile banks of the Waikato and Whanganui rivers. Some land in the north of the South Island was also farmed.

Tidy gardens

Soon after her arrival, Captain James Cook's *Endeavour* sailed north along the East Coast and stopped in one bay to examine Maori gardens. The ship's surgeon, William Brougham Monkhouse, wrote in his journal:

> We had an opportunity to examine their Cultivations more at leasure today and found them very far to surpass any idea we had formed of them. The ground is compleatly cleared of all weeds . . . with as much care as that of our best gardens. The Sweet potatoes [kumara] are set in distinct little molehills which ranged in some in straight lines, in others in quincunx. In one Plott I observed these hillocks, at their base, surrounded with dried grass.

A Cyprian Bridge watercolour of a Northland pa with extensive potato plantations. The potato was easier to grow and virtually replaced the fern root.

> The Arum [taro] is planted in little circular concaves, exactly in the manner our gardeners plant Melons . . . The yams are planted in like manner with the sweet potatoes: these cultivations are enclosed with a perfectly close pailing of reeds about twenty inches high. The natives are now at work compleating these fences. We saw a snare or two set upon the ground for some small animal [probaby kiore, the Polynesian rat] . . .
>
> It is agreed that there are a hundred acres of ground cultivated in this Bay – the soil is light and sandy in some parts – on the sides of the hills it is a black good mold. We saw some of their houses ornamented with gourd plants in flower. These with the Yams, sweet Potatoes and Arum are so far as we yet know, the whole of what they Cultivate.

Although many Maori continued to use the familiar digging stick long after the iron-bladed spade became available, the plough, introduced in 1820, became essential when large flat areas were to be cultivated.

Around the world, all populations had their own staple energy foods from grains such as rice, wheat, oats, millet and maize, and from root vegetables such as potatoes, yams, sweet potatoes, cassava and taro. In New Zealand, while Maori had access to native watercress (puha) and other vegetables, the food most commonly cultivated and prized was undoubtedly the tasty and nutritious kumara.

Like the populations of other farming cultures, Maori ritualised the cultivation of the kumara from the selection of the site of a garden, through preparation of the soil and planting in spring to a harvest festival style of digging the mature tubers in late autumn. Each step was accompanied by appropriate gestures and incantations from tohunga. The phases of the moon were taken into account, the stars observed, soil temperature tested and, among some hapu, a stone figure of the god Rongo was anointed where it stood in the field. Without care the tubers might be planted at the wrong time for early growth or harvested at the wrong time for storage. Church Missionary Society printer William Colenso wrote that the 'extensive tribal kumara plantations' were worked by tribal members 'in a compact body, chief and slave, keeping time with their songs, which they also sang in chorus'. After the intensive work of preparing and planting, the gardens would be tended in a more desultory way, but according to early Pakeha observers they were meticulously weeded and the hillocks hoed up throughout the late spring and the summer.

Before the main crop was dug late in March or in April, some of the larger immature tubers were fingered free from the underground stems. These were skinned, dried on mats, turned over in the sun and eaten as a late summer treat or stored. Sites such as Anaura Bay on the East Coast were ideal for farming, with the iwi able to avoid seasonal moves for other food. The soil and the climate were right and the sea and coastline fruitful.

The tohunga was the repository of knowledge, the man who knew the garden would last three seasons and then in its exhaustion be given over to fern root, the plentiful staple food that would grow there before it went back to manuka and could be burnt for a new kumara patch a few years later. The correct incantations and karakia (a prayer usually chanted) gave the Maori gardeners a sense of control over their destiny, invoking the spirits of their ancestors to intercede with the gods on their behalf. These invocations to the gods of food-growing, Rongo and Pani, were deeply poetic. William Colenso studied them but found translation extremely difficult because of allusiveness, ellipsis, metaphor and archaisms.

Taro and gourds were planted about the same time as kumara in the relatively few places where they could survive; the taro in rich, moist soil and carefully sheltered by brush fences against wind and cold. The leaves and stem, carefully cooked, were eaten as well as the root. Most gourds were lifted and their insipid flesh eaten but, as mentioned earlier, some were left to grow large and hard for use as calabashes.

Techniques for storing kumara, usually in deep subterranean pits, were developed over the years and thus gardens became increasingly larger and larger to produce food that could be safely stored to support hapu from one autumn to the next. A number of varieties developed in New Zealand in various soils and microclimates. The most common had a tuber described by early arrival Joel Samuel Polack as 'being the size and shape of the finger and extremely farinaceous and nutritious'. The large red varieties were brought in later by Europeans, in one recorded case by an American whaler in 1819.

Maori living in the northern coastal districts of the North Island had access to abundant food resources, provided by their own farming, fowling and fishing skills, even after the large land birds had been plundered to extinction.

A wider range of economic activity was practised by hapu in their villages than was once thought. For example, archaeologists investigated and partially restored a village they called Kohika, which was built on an island surrounded by a lake in the midst of what was once a large swamp on the Rangitaiki Plains. The village was a substantial settlement on the edge of the water, not unlike a modern marina, and thriving before its cataclysmic end not long before the arrival of the first Europeans. Its inhabitants gardened, gathered wild vegetables and took advantage of abundant fish and bird resources. The village was also a pivotal trading centre for wood and stone (mainly obsidian) because of its easy access to a network of inland waterways, to the sea, and to higher land. It was hierarchical in layout with a chiefly house in pride of place, and rich enough for some division of labour to allow for the specialisation of artisans. Its site was economically advantageous but also perilous and it was abandoned after a flood cut it off from its access to the lake, rivers and the sea.

Gardening tropical crops was never easy anywhere in New Zealand but the northerners learned to live with seasonal rhythm, planting, harvesting and storing, trapping birds and preserving any surplus in their own fat, catching eels, combing the foreshore for its rich resources, netting and hooking pelagic fish and drying them for storage. Green vegetables and the fruits of forest trees were readily available. As the naturalist Joseph Banks reported on the *Endeavour*'s visit: 'Tillage, weaving and the rest of the arts of peace are best known and most practised in the north eastern parts . . .'

Ethnologist David Simmons wrote of the spring and summer life of Maori at the time of Cook's first visit in 1769:

> At Anaura Bay, Wharetaewa and the Bay of Islands, the people were living in their base agriculture area either because of imminent attack . . . or because their access to sea resources was close and fruitful. In some cases then, the move to seasonal stations need not have yet taken place. This period in the agricultural cycle involved planting the crop, then living on fish and fern root which often also involved some form of dispersal to coastal areas or along coastal areas.

Rongo and Pani

Maori wove an elaborate and elegant mythology around the culture of the kumara, and the poems they spoke varied in style and content among iwi.

One of the supernatural offspring of Rangi the Sky Father and Papa the Earth Mother was Rongo-marae-roa (Rongo of the Great Domain) or Rongo-nui (Great Rongo). He was the god of peace and of cultivated foods, and often refined down to the god of the kumara. Rongo and Tane (God of the Forest) and Tangaroa (God of the Sea) are the most widely known among the deities of Polynesia.

According to another Maori myth, a relative of Rongo called Pani-tinaku (Pani the Germinator) gave birth to the kumara, and the different varieties of the plant that evolved here were named as the children of Pani. According to the pioneering scholar of Polynesian history and culture Elsdon Best, Pani also takes the form of a male in some other iwi myths, and in some she is married to Tiki, the male fertility figure.

Peace and agriculture were wrapped together in Maori society. Wars were seldom fought during planting and harvesting. Even during the civil war of the 1860s, Maori warriors supporting other hapu would go home to attend to their seasonal agricultural duties. The correct incantations and karakia invoked the spirits of their ancestors to intercede with the gods on their behalf for a magnificent crop.

The origins of the Polynesian kumara are obscure and somewhat mysterious. Most botanists agree it came from America, and this was a reason some anthropologists supported the now largely debunked claim that Polynesians themselves originated in South or Central America. The probability is that some Polynesians, after centuries of sailing eastwards into the sun, reached South America and found an intimidating land mass very different from the smaller, sea-girt islands of their homelands. Also, people already lived there and were easily able to repel prospective settlers arriving in canoes carrying no more than thirty or forty people. However, a few may have lingered long enough to carry the sweet potato and some other plants back to the Society Islands or the Marquesas.

The first explorers noted that Maori women had great expertise at dressing flax into a range of fabrics for everyday use, not only for clothing, baskets and blankets but also for the seine nets for fishing. This idealised picture of flax being dressed was published in the *Illustrated Sydney News* in 1865.

FLAX DRESSING BY THE MAORIES, NEW ZEALAND.

Simmons said that in Wharetaewa pa the people had a seine net 73.16 metres long, requiring a number of people to pull it. When conditions were suitable along the sandy beach the return on such a net would enable the people to stay in larger groups. Seine nets measuring more than a kilometre long were reported along the coastal stretches of the bays at Tokomaru, Anaura and Tolaga, where the beaches were sandy and sloping.

The abundance gave northern Maori the advantages of economic specialisation and more leisure to practise arts and crafts. Because the environment was relatively congenial, the highest concentrations of population were in the north, and by the time Europeans arrived this was imposing political pressures on land occupation and had increased the tempo of tribal conflict.

Life in the north of the South Island and on the shores of inland lakes and large rivers in the North Island may also have been relatively comfortable during the warmer cycle of years believed to have existed in the early times of Maori occupation, especially in the fifteenth century. But it became harder in the far south and inland regions of the North Island where the continental-type climate prevailed. When Europeans arrived the South Island coastal regions teemed with fish, seals and bird life, but other foods were harder to find. A botanist with Cook on his second voyage, George Forster, thought Cook Strait a miserable place, cold and swept by southerly and westerly winds, and populated by villagers who smelled of fish and rancid oil.

Cook left potatoes and other plants, and he and subsequent visitors, including the French explorers and whalers and sealers, continued to gift Maori of the southern temperate zone with plants and animals from the north. By the turn into the nineteenth century, the potato had replaced the kumara as the staple energy food. It could be planted and harvested between the planting and harvesting of the kumara as well as over a much wider region and a longer seasonal period. Maori were nothing if not pragmatic. Although some conservative groups continued to practise agriculture in the old way, most were bedazzled by the new plants, tools and animals. And settlers found that, by and large, the livestock and plant stock they brought with them fared better here than in Britain.

CHAPTER TWO

THE NEW CROPS, IMPLEMENTS AND ANIMALS

The weather was the main enemy of the Maori farmer, an enemy he fought with the rigorous canon of tapu, with elaborate litanies of hymn and incantation, but mostly with carefully acquired and memorised experience in coping with its caprices. Subsistence farming for a small population meant he was not seriously short of suitable land in the late pre-European period except in the coveted northern coastal regions. But when Pakeha arrived with iron implements, the potato, turnip, wheat, fruits and vegetables, pastoral and draught animals, and provided markets and a monetary transfer system, the weather became at once an ally and the infertility, the trees and steepness of so much of the land a commensurate liability.

As Maori embarked on something resembling commercial farming, their fields grew bigger and they became less scrupulous about rotating its use. They simply exhausted land and moved on. The competition for land and, to a lesser degree, competition for markets with produce, were factors in drawing the two competing races into war.

From the first, contact between the two cultures brought sharp social change to Maori. They had previously been occasional traders on a small scale – bartering kumara from the north for greenstone from the south, kumara and shellfish from the coast for birds and feathers from the inland forest. Obsidian and the work of artisans were also items of trade. But along the northern coasts of the North Island from the late eighteenth and early nineteenth centuries, Maori began to build their lives around the provisioning of ships and supplying the small settlements of whalers, sealers and entrepreneurs from Sydney.

The first sealers had arrived at Dusky Sound, only twenty-three years after Cook, in December 1792. Most Maori had long lived by the sea but there is evidence that more of them moved down to the points of the coast at which Pakeha regularly touched land to swap pigs, potatoes and flax for clothes and metal artefacts. The race was still in its fullest vigour, sharply intelligent, rudely practical and industrious. It would continue to flourish, despite disease and the rapacious incursions of musket-deadly war parties, until the early 1860s.

The halcyon decades of the Maori farmer were the 1840s and 1850s. They still cultivated the kumara in quantity but the potato was cheaper and easier to grow and had virtually replaced the fern root. Whereas the potato could be lifted from the soil after a relatively short growing season, eaten immediately or stored for long periods, the kumara had a longer growing season. Fern root needed tedious preparation before it could be eaten, and drying before it could be stored.

Their propensity for agriculture was given a fillip by the early missionaries, who communicated that farming was the purest efficacy of the work ethic – a measurable return for earnest labour. The arch-missionary Samuel Marsden was from Yorkshire farming stock and believed the maxim 'as ye sow so shall ye reap' to be literally true. He considered farming to be as pleasing to God as to man. He came to New Zealand to re-establish contact after

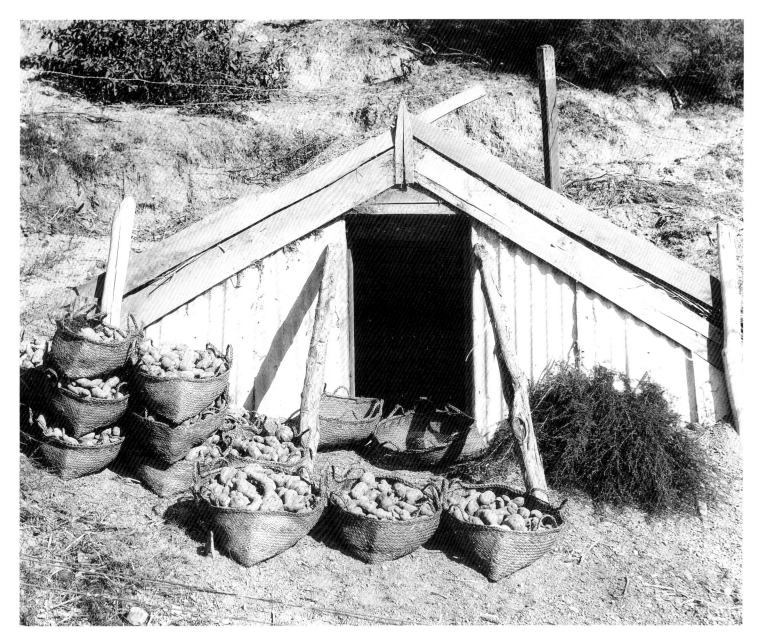

Maori storage techniques for kumara in pre-European times had led to an increase in leisure time which provided the chance for specialisation of labour. When the early Europeans came, Maori could grow large crops to feed the new arrivals.

the massacre of the crew of the *Boyd* in 1809 had scared off Australian and British interest in the country. Maori, with religious tensions surrounding agriculture, could easily understand what might be called the 'arable' parables of the Bible – if not yet the pastoral ones. From his first visit of 1814, the redoubtable Marsden encouraged the distribution of metal tools, so much more efficient than the wooden implements of the natives. He also brought with him the country's first horses, cattle and sheep, accompanied by men with some knowledge of stockmanship.

But if the missionaries were enthusiastic about encouraging Maori in agriculture, how able they were to instruct them in the task is debatable. In their centennial publication *The Farmer in New Zealand*, G. T. Alley and D. O. W. Hall wrote:

'The missionaries, although they vowed the intention of instructing the Maoris in the arts of civilization, were not themselves masters of the arts they taught – not at least of the art of farming . . .'

Those missionaries, though, were generally a tough and indefatigable cult of men, hewn from the granite of Anglo-Scottish evangelism. A Ngapuhi chief called Ruatara, who had been round the world in European vessels, had spent time on Marsden's farm at Parramatta in New South Wales. According to Marsden, Ruatara returned to the Bay of Islands and 'laid out the grounds he intended to clear and cultivate, and marked out the ground for his men, having first enquired of me how much ground a man broke up in a day at Port Jackson'. Marsden held out great hope for Ruatara as a progressive influence on Maori and their personal attachment

An idealised engraving by Samuel Williams in 1836 of the Church Missionary Society's mission station at Waimate North. Missionaries were enthusiastic about encouraging Maori agriculture but how able they were themselves is debatable.

seemed strong. The missionary was deeply saddened when Ruatara died suddenly from a mysterious illness soon after Marsden's return to Sydney from his first visit to the Bay of Islands in 1814. The tough and able Hongi Hika, also a former visitor to Parramatta, took over the leadership and organisation of Maori agriculture in the north – but ultimately found warfare much more rewarding and his musket-toting warriors became the scourge of the North Island in the 1820s.

Maori quickly became proficient farmers of the new plants at their disposal. The first Pakeha visitors were fed by them. By 1830 the natives were involved in export trade. In that year twenty-eight ships (averaging 110 tonnes) made fifty-six voyages between Sydney and New Zealand. Maori-grown potatoes were the cargo on those ships, soon to be followed by Maori-grown and -milled grain. Also that year, an agricultural mission school was set up in Waimate North, almost twenty kilometres inland from Paihia. Five years later it was visited by naturalist Charles Darwin, who wrote: 'On an adjoining slope fine crops of barley and wheat were standing in full ear; and in another part fields of potatoes and clover . . . There were large gardens with every fruit and vegetable which England produces . . .'

However, not all the action was in Northland. In his history of the East Coast region, *Challenge and Response*, W. H. Oliver wrote: 'The economy of the Maoris changed as they began to produce for a market. First it was potatoes and pork for visiting ships, then those foodstuffs and flax for visiting traders. The missionaries, beginning in the later 1830s, brought, among other things, improved agricultural techniques. The Maoris next learned to build and run European-type vessels.'

Oliver says the wheat boom of the 1850s, the Maori-operated export trade to Auckland from the East Coast, and the quick realisation that the land was an economic asset, all derived indirectly from Hongi and the rampaging Ngapuhi: 'Christianity reached the region ahead of its regular apostles,

Above: John Morgan took over the Church Missionary Society's station at Otawhao (Te Awamutu) in 1841, under the protection of Rewi Maniapoto. It was because of Morgan's enthusiastic guidance that the region blossomed as a Maori agricultural centre during the late 1840s and the 1850s.

Right: This picture of Kororareka in the Bay of Islands demonstrates that Maori vessels were heavily involved in trade.

carried by Maoris enslaved by the Ngapuhi, and freed after they had accepted the new religion. These unofficial new agents first spread the new ways; the missionaries came to take over a going concern.'

And with some of the missionaries came a measure of agricultural expertise. By the early 1840s, William Williams had introduced wheat-growing to Waiapu and Maori were growing and milling the grain and exporting it to Auckland.

Gisborne historian J. A. Mackay says that well before wheat was cultivated in Poverty Bay maize had been introduced and was being extensively grown by the 1830s. In the Waikato, missionary John Morgan was involved in the brief flowering of Maori agriculture during the two decades before the civil war, after which Pakeha snatched much of the best agricultural land for their encroaching compatriots and destroyed the sanguineness of the aboriginal inhabitants.

By all accounts, and there were many, the era of peaceful progress and industry among Waikato and Ngati Maniapoto was from about 1845 to 1860. Then the Taranaki War, the forerunner of the Waikato campaign, interrupted the new and profitable era of productivity, especially wheat-growing and flour-milling, which preceded brisk trade in

The rider was 'more than mortal'

Samuel Marsden's first visit to New Zealand in December 1814 was even more significant for its effect on Maori agriculture than on the future of Christianity in this country. Marsden set out from Sydney in the mission ship *Active* on 28 November with a party of thirty-five, including missionaries Thomas Kendall, William Hall and John King, eight Maori, two Tahitians – and a runaway convict who had stowed away and was discovered at sea. However, it was the cargo of livestock that was to make the biggest impact, listed as 'one entire horse, two mares, one bull and two cows with a few sheep and poultry of different kinds'.

Among the passengers on the *Active* was John Liddiard Nicholas, a prolific writer, who said that when the animals were landed at Rangihoua, seven miles from the heads in the Bay of Islands, Maori were bewildered and amazed:

Their excitement was soon turned into alarm and confusion, for one of the cows that was wild and unmanageable, being impatient of restraint . . . rushed in among them and caused such violent terror through the whole assemblage that, imagining some preternatural monster had been let loose to destroy them, they all immediately betook themselves to flight.

When Marsden mounted a horse and rode up and down the beach they followed him with staring eyes, believing for the moment he was 'more than mortal'.

Thirty-four years later, horses and the pastoral animals had multiplied many times. A census in Auckland was taken and stock numbers in the district alone were: cattle, 137,204; sheep, 1,523,324; horses, 14,192; mules and asses, 122; goats, 11,797; pigs, 40,734.

A wood engraving of a busy, peaceful Taranaki pa, which appeared in the *London Illustrated News* in 1860, about the time the war broke out in the province.

Auckland. James Cowan, a writer brought up on a Waikato farm on land confiscated after the civil war, said the heavy work and the harvesting on Maori farms were done by means of an ohu (working bee). 'Waikato's surplus was taken to Auckland in large canoes,' he wrote. 'This consisted chiefly of wheat, flour, maize, pigs, fruit and muka (dressed flax). After each harvest the Waipa and Waikato rivers were busy with this canoe transport going down fully laden to the market and returning with goods purchased in Auckland. The route was via Waiuku, which was then a very busy place, and Onehunga.'

Wheat was first introduced to New Zealand by Jean François Marie de Surville of the *Saint Jean Baptiste* the same year Cook first arrived. He presented the natives with some wheat and other seeds and tried to explain how to cultivate them. Then Marsden gave Ruatara some seed at Parramatta and he brought it back to the Bay of Islands trumpeting that this was the plant from which the Pakeha made his biscuits. The wheat grew strongly, but at first some Maori, unfamiliar with grain, dug it up as it began to dry to find the food at the root. Ruatara and Hongi allowed the grain to ripen, threshed it, milled it and made a cake to convince their countrymen. That was in 1813. The wheat strain was of Indian origin via New South Wales.

The wet and humid summers would always prove a problem for wheat-growing in the north and quite early it was discovered it did better on the drier eastern plains of the South Island. A settler, according to F. W. Hilgendorf's *Wheat in New Zealand*, reported that in the 1840s, Maori on Banks Peninsula were familiar with wheat:

'Man was the first horse'

The first plough was brought in by pioneer Anglican missionary John Butler and used at Kerikeri. A Marsden protégé, Butler wrote sanctimoniously in his diary:

> On the morning of Wednesday the 3rd of May 1820, the agricultural plough was for the first time put into the land of New Zealand at Kiddie Kiddie [Kerikeri] and I felt much pleasure in holding it after a team of six bullocks brought down by the *Dromedary*. I trust that this auspicious day will be remembered with gratitude and its anniversary kept by ages yet unborn. Every heart seemed to rejoice in the occasion. I hope it will still continue to increase and in a short time produce an abundant harvest.

Well, it was auspicious. In the halcyon days of wheat-growing in the Waikato two decades later, Maori had many ploughs drawn by pairs of horses. Yet before they could afford horses or European-made ploughs they had made their own copies entirely of wood, which were drawn by men. As an old man told writer James Cowan: 'Ko te tangata te hoiho tuatahi' (Man was the first horse).

Although the early missionaries in the Bay of Islands were not expert farmers, one immigrant soon arrived who turned the Waimate mission property into the first fully mixed farm after the English model, with both crops and livestock. He was Richard Davis, from Dorset, and in 1831 he took over the 250 acres bought from local Maori. He was a serious and efficient farmer who made an immediate success of the mission operation, which declined after he moved on.

Although Butler didn't stay long in New Zealand he was an agent of significant change. He had the previous year been a signatory, with fellow missionary Thomas Kendall, to the deed which recorded the first land sale in New Zealand. It was 4 November 1819 and the Church Missionary Society paid forty-eight 'falling axes' for 13,000 acres of land at Kerikeri from 'Shunghee Heeka' (Hongi Hika). Butler and Kendall had set in motion the land rush.

The double-tine plough began to appear during the 1870s and farmers graduated to three and sometimes four blades as soon as they had the bullock or horse power to do so. The first tractors appeared about 1910, without making a great impact, sometimes because of unreliability. World War One research and experimentation with vehicles saw a leap in their development and from the 1920s their use became widespread, especially among those making a living from cropping. The extra horsepower meant more tines and thus more furrows at a run. The tractor and plough were also designed so the implement could be controlled by the driver without dismounting from the vehicle. The tractor thus greatly extended the amount of land able to be ploughed in a day by one man.

Above: Anglican missionary John Butler introduced the plough to New Zealand and hoped the anniversary of 3 May 1820 would be remembered with gratitude 'by ages yet unborn'.
Left: Maori men pulled their own ploughs at first, but then bullocks took over and after that draught horses and then tractors.

'When digging potatoes they sowed the grain as they dug, and in turning over the soil covered the grain. In this way they would sow as much grain as was covered by a day's digging. The crop, of course, did not ripen evenly but as they cut it with a hook that didn't matter. They cut it as it ripened.'

According to John Morgan, so much wheat was grown at Rangiaowhia, near Te Awamutu, by the summer of 1845–46 that he arranged the construction of a water-powered flour mill. Maori were at first only mildly interested but by the time the mill was finished and functioning, in 1848, the benefits were clear and many hapu around the upper Waikato area wanted mills of their own. Before the turn of the decade there were 450 acres of wheat supporting the mill at Rangiaowhia. 'I had also introduced barley and oats,' wrote Morgan. 'Many of the people at various villages are now forming orchards, and they possess many hundreds of trees budded or grafted by themselves, consisting of peach, pear, plum, quince and almond; and also gooseberry bushes in abundance. For flowers or ornamental trees they have no taste; as they do not bear fruit, it is, in their opinion, loss of time to cultivate them.' The peach trees seemed particularly to thrive. The fruit was sliced and strung up in the sun and wind to dry for use in winter pies and tarts. It was so abundant it was also used to fatten pigs and even feed horses and cattle.

Austrian geologist Dr Ferdinand von Hochstetter wrote of a visit to the upper Waikato in 1859: 'The beautiful, richly-cultivated country around

By the end of the golden age of Maori farming, after the wars broke out in the 1860s, many flour mills existed to handle the large areas of wheat. Similar to this mill, built in 1850 at Waimate North, they were driven by large water-wheels with this sort of gearing.

Rangiaowhia and Otawhao [Te Awamutu] lay spread out before us like a map. I counted ten small lakes and ponds scattered about the plains. The church steeples of three places were seen rising among orchards and fields. Verily I could hardly realise that I was in the interior of New Zealand.'

But for bucolic bliss, nothing beats the description of a Maori farming area in the Waikato in the early 1850s by Lady Martin, wife of the first Chief Justice:

> Our path lay across a wide plain, and our eyes were gladdened on all sides by sights of peaceful industry. For miles we saw one great wheat field. The blade was just showing, of a vivid green, and all along the way, on either side, were peach trees in full blossom. Carts were driven to and from the mill by their native owners, the women sat under the trees sewing flour bags; fat, healthy children and babies swarmed around, presenting a floury appearance . . . We little dreamed that in ten years the peaceful industry of the whole district would cease and the land become a desert through our unhappy war.

Another, more fact-rich, description of Maori farming came from a member of the Legislative Council, William Swainson, who wrote that a native population of about eight thousand in the Bay of Plenty, Taupo and Rotorua, in 1857

> had upwards of three thousand acres of land in wheat, three thousand acres in potatoes, nearly two thousand acres in maize, and upwards of a thousand acres planted with kumara. They owned nearly a hundred horses, two hundred head of cattle and five thousand pigs, four water-mills and ninety-six ploughs. They were also the owners of forty-three small coasting vessels, averaging twenty tons each, and upwards of nine hundred canoes.

Swainson, a numbers man, reported that in 1858 at the port of Auckland (Waitemata Harbour) fifty-three small vessels were registered as being in native ownership, and the annual total of canoes entering the harbour was more than seventeen hundred.

On 7 March 1848 Governor Sir George Grey conveyed in flat, rather patronising prose in a letter to the Secretary of State for the Colonies, Earl Grey:

> It will be interesting to your Lordship to be informed that during my journey through the extensive and fertile districts of the Waikato and Waipa I was both surprised and gratified at the rapid advances in civilisation

With thousands of hectares of wheat, potatoes, maize and kumara, and thousands of pigs, Maori were in a strong position to trade with Pakeha and were tough and adept at bartering. This picture shows traders haggling with Maori up-river from Wanganui.

which the natives of that part of New Zealand have made during the last two years. Two flour mills have already been constructed at their sole cost and another water-mill is in the course of erection. The natives of that district grow wheat extensively: at one place alone the estimated extent of land under wheat is a thousand acres. They have also good orchards with fruit trees of the best kind, grafted and budded by themselves. They have extensive cultivations of Indian corn, potatoes, etc., and they have acquired a considerable number of horned stock. Altogether I have never seen a more thriving or contented population in any part of the world . . .

An anonymous writer in the *Maori Messenger* the following year described 'the singular energy and ability' of Maori in competing with Europeans in farming and other industries as 'by no means one of the least remarkable events of the age we live in'.

Auckland fed and flourished on the food from Maori canoes, which returned to their villages with implements, clothing and much-sought-after European novelties. Maori proved shrewd negotiators but did not understand the cyclical nature of capitalist economics. An agricultural depression at the end of the 1850s saw their affluence shrivel somewhat. They had come into direct confrontation with Pakeha both as farmer and provider. In 1844 Governor FitzRoy wrote: 'The natives are well inclined to labour for very small remuneration.' He saw this as an advantage, but others saw it differently. As early as 1845, one commentator said Maori produced food so cheaply they would ruin the white settlers' market. Their success as farmers contributed to their undoing. As Alley and Hall wrote: 'The white settler had his own place for the Maori in the scheme of things, and it was certainly not that of a competitor. The Maori was expected rather to be the willing accomplice in converting his forests into the white man's farms.'

Keith Sinclair in his *History of New Zealand* wrote, 'In agriculture the Maoris, who competed directly with the settlers, enjoyed considerable advantages. They owned most of the good land in the North Island. They farmed communally, and thus had no labour costs. Until the 1860s they probably produced the greater part of the food crops consumed in New Zealand or exported to Australia.'

Charles Barraud (1822–97) jotted impressions in his sketchbook of Maori life after European settlement, many of which were later published in London.

Depressed prices and then the Taranaki and Waikato wars brought to an abrupt end the Maori adaptation to temperate farming in the context of a market economy – an adaptation that was an extraordinary feat measured against the performance of other New World races facing the onslaught of European discovery and expansion. The confident pride with which Maori had farmed in Northland, the Waikato, Bay of Plenty, King Country, Taranaki and the Horowhenua district in the 1840s and 1850s and fed Pakeha settlers was long neglected by New Zealand historians. The tone of conventional history was for so long that the land was the richer for being taken over by Pakeha because Maori would leave it idle at worst or farm it for subsistence at best. But Maori mixed farmers had adapted brilliantly in many places until the pressure from politicians, speculators and Pakeha settlers began forcing them off the last of their land that was most suitable for arable and pastoral farming.

Maori fought back too late, outnumbered and underequipped against a better-organised foe, and depressed by a growing understanding of the

Implements and machines

The digging stick, or ko, was the most frequently used gardening implement by pre-European Maori. It was similar to those used in many other parts of the world where iron and steel were either unknown or not readily available, and especially resembles the 'Highland spade' still used in Scotland into the nineteenth century.

Most Polynesian populations had a version of the ko, or something very close. In New Zealand it was used throughout the country, varying in size and style but generally made of maire or matai, hardened by ageing or fire. It was around two metres long with a sharpened blade eight to ten centimetres wide at the bottom. Those used for turning over the light sandy soil before planting kumara had a footrest lashed into a slot on the side of the blade so diggers could use foot power to help thrust the implement deeply into the ground. Many were carefully made and decorated with carvings at the handle end. Similar implements, but without the footrests, were used to dig post-holes for palisades and houses. Smaller sticks, some shaped like paddles and most commonly called kaheru, were used by workers either sitting or hunkering to break up clods turned over by the ko.

Iron spades were immediately in great demand among Maori, especially for constructing earthworks to protect pa; but the ko lasted many years among gardeners, even after the potato became the most prolific crop. Just as Maori were pragmatic in their approach to the availability of new tools, all New Zealand farmers generally adopted mechanisation as soon as they had the capital to purchase.

The first cereal crops in New Zealand were harvested with sickles and scythes. The first mechanical mowers cut swathes along the paddocks in the 1870s but the sheaving was still done by hand for another decade when reapers and binders made their appearance and steadily developed in capacity and speed. The equipment encouraged arable farmers to cultivate more carefully to reduce the number of weeds harvested by the unselective machines. Tractors gradually took over the task of pulling the machines but horses still commonly did the job until well into the first third of the twentieth century. Header-harvesters appeared not long before World War Two.

Labour-saving mechanisation is so attractive it can change farming practices, even if only temporarily, for no other reason that it makes life easier. For example, silage grew in popularity in the early decades of last century, until 1932–33 when 116,000 acres were harvested, according to Department of Agriculture records. But the harvest declined over the following decade to 53,000 acres in 1943–44. The reason was increased mechanisation of haymaking, which cut labour and speeded up the harvest, thus reducing the danger of being caught by the weather. Silage made a comeback against hay but the most important trend was towards working off the living leaf – with careful animal and pasture management. Autumn grass, for example, was saved for the pinch period before spring struck.

Shearing machines became available in the first decade of last century, quite quickly replacing blade shears except in high country

Shearers used an early Lister system made in England, but employed here and in Australia on a much larger scale, working on much larger flocks.

cynical, acquisitive forces ranged against them. In selling much of their land, Maori had succumbed to the blandishments of some missionaries and of politicians. Finally, when Pakeha stratagems would not work any more, military force was applied. Wrote Alley and Hall in 1940:

> [T]he British army officers engaged in the campaigns (the British troops were withdrawn gradually after the subjugation of the Waikato) were wont to accuse the colonists of prolonging the wars for the sake of their pockets.

Certainly the market for fat stock was vastly improved by army purchases. The northern settlers benefited in another way. The destruction of the military power of the Maori meant their elimination as competitors in food production, a result partly due to the voluntary retirement of the Maoris after the wars.

A sad image remains: Rewi's Last Stand at Orakau was fought in a peach grove on the edge of where, fifteen years before, wheat fields had held such promise for a Maori farming future.

where sheep would suffer from too close a shave. The first source of power was a portable steam engine, which was replaced by internal combustion systems and then by electric power as reticulation spread through the rural districts in the 1920s and 1930s.

Starting in the late 1880s, steam-operated centrifugal separators launched the factory system for making butter and cheese. Local creameries were set up which enabled farmers to take their milk each morning, retrieve the skim milk, and leave the cream for butter-making either at the creamery or at a central dairy factory.

The mechanical milking machine was a huge boon to the dairy farmer, releasing some of the family from the grind of twice-daily hand-milking. Family units remained economic as herd sizes increased. A number of types of machine came quickly on the market during the first two decades of last century, some imported and some developed locally. Initial opposition arose. The Department of Agriculture annual report in 1906 was pleased at the spread of the machine but added:

> If the dairy experts can assure dairymen that the use of the machines will not damage the teats or interfere with the milking qualities of the cows, and that the milk is not deteriorated for butter or cheese-making, then the general introduction of the machines will only be a matter of time, and a great impetus will be given to the industry.

Once the number of cows in a herd passed twenty, machines became essential. Some of the early teat-cup liners were rigid, but over the years they became increasingly soft and flexible. By 1920 nearly 9000 farms had milking machines and this figure trebled over the following fifteen years.

As with shearing machines, the first milkers were operated by steam engines, then internal combustion oil engines and, later, electricity when it became available. The number of electrical systems for milking machines and separators passed the number driven by steam engines in 1930, but five years later two-thirds of the machinery was still using internal combustion engines. A small minority of farmers, presumably those with small herds, still milked by hand at the beginning of the 1950s. The first machines collected the milk from each cow into a bucket which was poured into a larger vessel. These machines were replaced by a system which used a vacuum tank to draw the milk from all the teat-cups into a single tank ready to be delivered to the separator.

The benefit of refrigeration to the dairy industry did not stop with the ability to export products to other countries. Refrigeration plants in dairy factories allowed temperature control to be exercised over the manufacturing process. Achieving a consistent standard of butter quality for export was difficult enough for more than twenty years after the export trade began because of primitive and often unhygienic conditions on undercapitalised dairy farms and in some small and badly run factories. However, as temperature control systems spread through the factories and ultimately back to the farms, high-grade butter became the norm.

Early milking machines pumped the milk from one, or sometimes two, cows into a small container which then had to be poured into a larger milk can. Later the milk was pumped from each cow directly into the one large receptacle.

CHAPTER THREE

A LANKY LAND

European settlers arriving in New Zealand found a lanky land running from south-west to north-east for 1600 kilometres, warmly temperate around the coasts but just wide and high enough in the central regions of both islands to produce something approaching a continental climate, with cold, dry winters and hot, dry summers. Mostly the evergreen seasons of this land move from cool and moist to warm and moist and back again without the joy of a Hey presto! spring or the gloom of a cloistral winter.

Original Maori immigrants would have found the climate harsh in contrast to what they were accustomed to but for the Europeans it was benign. In almost all the farmed areas rainfall varies between seventy-five and 175 centimetres from showers on something like 160 days a year, with between 1800 and 2400 hours of sunshine. The frequent alternation of rain and sunshine distinguishes the New Zealand climate from other temperate zone conditions around the world. The prevailing wind is north-westerly and the country is a windy one. The earliest European settlers soon discovered the advantage such a climate bestowed on their type of farmer – namely, that in most regions pastoral animals could be left outside on grass that grew pretty well year-round, without the need for expensive sheltering buildings. Shelter belts of trees did the job.

A relief map of the country is a busy picture. A mountain axis runs from Fiordland up the western side of the South Island, through the south-west tip of the North Island to East Cape. Only about a quarter of the 26.9 million hectares are under the one hundred metres contour. Broadly speaking, half the country is mountainous, about a quarter of it hilly, and the rest rolling or flat.

The West Coast of the South Island and a strip along the south-east coast, and almost all the North Island were covered in heavy bush when the first Europeans arrived. The type of forest varied with altitude and geographic situation but fell into two categories. One was rainforest in the North Island and some parts of the South Island, such as Westland and Southland, generally at lower altitudes, with tall evergreen trees as a top storey, other lesser trees of many species and tall ferns in the middle, and a heavy latticework of climbers running up from luxuriant undergrowth on the floor. It was difficult forest to penetrate, let alone raze and develop into pasture. The other was beech forest, generally confined to the higher altitudes, running along the foothills, crouched against the mountain spine of both islands.

The flat and rolling country of the South Island is clear-cut enough, with a pocket at Nelson, a small plain in Marlborough, a strip along the coast of Westland, the big stretch of the Canterbury Plains, and on down through eastern Otago and round the bottom of the island into the plain of central Southland. The high country that is farmed runs up the eastern side of the Southern Alps with grazing almost to the snowline in summer and has hill-country ridges extending to the coast in North Canterbury, Otago and Southland.

In Nelson and Marlborough the soils are recent

alluvials in yellow-grey and yellow-brown earths. Quite early, they were supporting market gardens, orchards and specialist crops as well as sheep and a few dairy cows. Large herds of beef cattle swarmed the high, cold rangeland of inland Marlborough. Nelson is a kind of temperate Eden with an average of 2500-plus hours of sunshine each year and between 1200 and 1500 millimetres of rain which, like mercy, is not strained and droppeth gently over an average of 120 days. This salubrious region was, ironically, the source of major ingredients of the country's cigarette and liquor production – Nelson grew all New Zealand's hops and tobacco – until tobacco was phased out during the 1990s. It continues to grow all the hops.

Grey podzols with recent swamp soils on alluvial flats support the farming on the narrow West Coast plain, which has become increasingly important as a dairying area, particularly in the south. Canterbury soils range from stony gravel to fine silts with high natural fertility, and there are extensive deposits of gravel and silt on the Southland Plain. Otago's back country has yellow-brown earths on the hills and yellow-grey earths, often stony, in the basins.

By the turn of the twentieth century stock had been carried for forty years on accessible areas of easily developed pastoral land carrying tussock on the eastern side of the South Island, the Wairarapa and southern Hawke's Bay. The only other tussock country was less accessible – a pocket in Poverty Bay and a large open-range area east and south of Lake Taupo which, it turned out, had a cobalt deficiency that affected stock health. But as the century turned the signs were there – that in the twentieth, despite the heavy counterpane of forest, the North Island would wrest from the South the economic superiority its agriculture had enjoyed in the nineteenth.

The major farming regions of flat and undulating land in the North Island are listed below.

The Waikato-Thames-Hauraki Plains are composed mainly of volcanic sediments derived from the Central Plateau. The plains of the North Island are usually composed of finer sediments than those of the South Island. A considerable portion of the area was formerly swamp; but drained, it is now good farming land. The Waikato-Thames-Hauraki

Plains are separated by a low range from the small, disconnected plains of the Bay of Plenty.

The Poverty Bay Flats are the valley plain of the Waipaoa River. North and south of this plain, discontinuous patches of flat land occur in the valleys of minor streams and along coastal strips.

The Heretaunga Plain is the most extensive area of flat land on the east coast. It is roughly triangular and is formed of alluvial sediments deposited by the rivers which cross the plain. After the earthquake of 1931 land was reclaimed from the Ahuriri Lagoon, near Napier.

The Taranaki Plain is an elevated plain ending at the sea in cliffs. Its surface is broken by the courses of rivers which emerge from the hill country. Thus there are small areas of river-terrace plains lying at a lower level than the general elevated surface of the country.

Hereford and red Devons wintering in the Wairarapa. Early European farmers found that New Zealand's mild climate meant that in most regions cattle could be left outside through the winter to graze, with some supplementary feed.

The Manawatu-Horowhenua Coastal Plain stretches from the Manawatu Gorge to the coast. From the gorge to Palmerston North the plain slopes steeply, but between the town and the sea the slope is more gradual and the river follows a winding course across it. This part of the plain is separated from the sea by a line of sand dunes. The rainfall on the coastal belt is sufficient for dairy farming on the light sandy soil. Southward the coastal plain becomes narrower and is then known as the Horowhenua Coastal Plain, as it is largely contained in the Horowhenua county. This plain consists of a series of terraces along the foot of the high country, while on its outer edge it is fringed by sand dunes and swampy hollows. These hollows, which are numerous in the vicinity of Levin, were formerly covered with flax. Foxton was once the centre of the flax industry in the nineteenth and early twentieth centuries.

The Wairarapa Valley runs parallel to the high country and separates it from the coastal hill country. In its upper portion the plain slopes steeply, but southward it widens and the slope is more gradual, and much of the land is good for dairying.

In the North Island hill country, laboriously farmed with sheep, the major regions are listed below.

The North Auckland Peninsula is complex in character. The chief areas of flat land are those along the rivers in the coastal areas. The region is mainly a soft-rock upland of low elevation, generally below 200 metres, but with several peaks rising to 300 and 600 metres. The eastern coastline of the peninsula is often irregular, with deep inlets, extensive estuaries and mud flats, while on the west coast there are long stretches of sand dunes, notably along Ninety Mile Beach in the Far North.

The Central Volcanic Plateau occupies the greater portion of the middle of the North Island. The main systems of river drainage radiate outward. The plateau is built of volcanic material, its surface being relatively even, but often broken by volcanic cones and irregular ridges. Large areas have been used for the establishment of exotic forests, and farming has been successfully established.

Inland Wanganui-Taranaki Hill Country fringes the western boundary of the volcanic plateau from which it stretches to the volcanic cone of Mt Taranaki/Egmont. The hill country surrounding it is composed of soft rocks. The highest ridges are those further inland and overlooking the volcanic plateau, rising in the east to 900 metres. From the highest ridges there is an even slope of the land towards the coast. The whole plateau has been intricately and deeply dissected by streams. The ridges between the stream valleys have sharp crests and steep sides. This area of weak rocks and steep hill slopes is also one of high rainfall, where formerly the soil was protected by a dense forest cover. Much of this type of land has been cleared and used for grazing stock; today some areas are reverting to second growth, while there is some soil erosion on the steeper slopes.

The East Coast Ranges and Hill Country extend about 550 kilometres, from Cape Palliser to East Cape. Throughout the southern section the ranges are narrower than they are to the north of the point where they are cut by the Manawatu River. From this river gorge they widen out, with the Kaimanawa Range extending westward towards the volcanic plateau as a wide belt of rugged mountain country. On the eastern side of the main range there is an extensive belt of hill country stretching to the coast both to the north and south of Hawke Bay. The northern ridges of this soft-rock hill country lose elevation seaward. The slopes facing inland are steep, while larger and more gradual slopes face the coast. Few large rivers run through these valleys and it is only in the lower reaches that there are any areas of flat land. To the south of Hawke Bay the main ranges are separated from the coastal hill country by low rolling land or by wide gravel valleys.

North Island soils have been strongly influenced by the fallout of volcanic ash, with surface deposits found as far south as Wellington and as far north as Kaikohe. But the area dominated by volcanic ash soils is the centre of the island, from Wanganui to the Waikato Heads on the west and from Napier to Whitianga on the other coast. In the centre and along the east coast the predominantly yellow-brown pumice soils are low in elements, notably phosphorus, potassium, magnesium, cobalt, copper, iron and boron. To the north, west and south the

Nelson hops have an international reputation, with some of the most famous breweries in the world – Guinness, for example – regularly using them. The plant is particularly site-specific. It is grown successfully in few parts of the world because it is sensitive to latitude (hours of daylight), to sunshine hours and to the spread of rainfall, and requires protection from wind. Fewer than thirty growers produce about 800 tonnes a year, eighty-five per cent of which is exported.

A Hawke's Bay coastal property at Ocean Beach, south of Cape Kidnappers.

older volcanic ash has formed yellow-brown loams, less deficient in trace elements.

That is how an agriculturist sees the land that is New Zealand, but it does not fully describe the difference between the two main islands – that in the South there is a different scale of distance and size, longer flats and higher mountains, with bold, faraway horizons that allow a special kind of reflective patience to overlay a person's thinking. The southern pioneering farmers had time and space and few obstacles to overcome, so their development was predictable and sure. They were steady and tough and by temperament conservative, and as the end of the nineteenth century approached they seemed to have run out of drive. And debt was an encroaching problem.

In the North Island the landscape is newer, smaller in scale, with a close and shifting horizon bringing surprise with every twist of a journey. Northern settlers found this broken countryside with its endless walls of trees a dogged adversary that gave them little time to think during the thirty years from 1880, only time to slog at the forest face, cutting and burning and sowing grass in warm ashes – turn their backs on the bush and it writhed remorselessly through the grass again.

They took on the bush and won. But it was a long, hard fight.

Right: Pastoral farming has been successfully established on the North Island's central volcanic plateau since a cobalt deficiency in the soil was overcome, but the region has been most important for the establishment since early last century of a sustainable exotic forestry industry on which pulp and paper production was based.

Opposite: Under the shadow of Te Mata peak, near Havelock North.

CHAPTER FOUR

THE FIRST EUROPEAN FARMERS

The next phase in the farming of New Zealand was large-scale wool-growing, an export-orientated enterprise that brought this country its first taste of affluence. It developed in the south of the North Island at the time the zenith was approaching in the north for the Maori-dominated mixed agriculture that fed the earliest settlers and opened the export trade in food. The first pastoralists were from Wellington – the least likely town, locked off as it was by mountains and sea from the grassed valley flats or undulating tussockland they needed for sheep. These men got the idea – and then the sheep – from Australia, where fortunes were being made from merino wool.

Wellington juts its jaw out over the edge of the South Island, and is often hit by a southerly uppercut brought up from the floor of the world. The original idea was to build the city in the Hutt Valley, on the present site of Petone. However, the first settlers in what was called Britannia were flooded out in 1841 by the Hutt River and moved on to the Te Aro flat and the hills of Wellington. This disqualified the valley as a possible sheep run, for the time being anyway. Soon, the New Zealand Company's close-order, tidy style of settlement began to fail for want of suitable land to farm; and local Maori disputed ownership of what there was.

Life was hard for Britain's working class and lower middle class at the time, and especially for rural workers. The steadily increasing enclosure of common land denied them their subsistence farming. The Industrial Revolution was driving them into the towns and cities, where poverty was worsened by the immense population growth during the nineteenth century. A driving force behind the enclosures was an agricultural revolution that began in the late eighteenth century. Animal and plant breeding and planting techniques were improving and mechanisation was increasing, reducing the demand for labour and providing an incentive for wealthy landowners to fence off their own land and then common land. They did so with little thought for the chaos that would ensue. Unemployment led to pauperism and then to spreading lawlessness in the densely packed cities, and emigration was seen as an answer. Edward Gibbon Wakefield and his New Zealand Company associates had theorised the development of a colony, using the village structure of self-sufficiency gradually spreading outwards, selling land to settlers and using the payments to bring out more migrants and to purchase additional land.

The scheme was not working in Wellington, which was by 1842 an unhappy town snuggled in bush that made the land difficult and expensive to get into production. As Joseph Masters, the man who helped found Masterton, wrote in a memoir (published by The Wairarapa Archive in 2005): 'The only land available around Wellington at this time [the late 1840s] was the wooded hills around Wellington, totally unfit for agricultural purposes.' The small-farmer concept was vested with much Protestant virtue. It was considered the right sort of thing to do as much in moral as in economic terms. Alley and Hall noted, 'In Otago the *News*, within a short time of the beginning of settlement,

HERE AND THERE;
OR, EMIGRATION A REMEDY.

In the 1840s and 1850s life was hard for working-class and lower middle-class Britons, especially those from rural areas driven into the cities after the enclosing of common land deprived them of their traditional subsistence farming. Those who stayed in the country were also under pressure, with few farm jobs available. This was a time of dramatic population expansion. The British Government was happy for emigration opportunities to absorb its surplus people as poverty and lawlessness grew, and New Zealand and other colonies were eager to boost immigration, as the pictured cartoon from *Punch* dramatically shows.

repudiated the idea of agriculture altogether, advocating the development of pastoral farming as the only road to prosperity, an advocacy that so angered the citizens of Dunedin that the paper came to an end late in 1850 for lack of influential support.'

It became obvious to one group of Englishmen in Wellington, ensnared only briefly by the New Zealand Company dream, that a better way for a man with capital to get into fast money was to find grassland and graze sheep. The Australian experience was there as evidence. Wool was the only commodity that could foot it with timber as an export. If arable farming had about it a work-ethic air of wholesomeness for the working class and lower middle class, sheep farming was glaringly a business enterprise that appealed to the instincts of the upper-middle-class Englishman. So these men, some of them relatives, formed partnerships to invade the Wairarapa Valley. One of them, a future premier, Frederick Weld, according to his biographer, said: 'I never had any taste for pottering about with little bush cultivations, an occupation more suitable to labourers with large families.'

The New Zealand Company was a driving force behind the early emigration from Britain to the colony and, although its scheme of orderly settlement didn't work in detail, all the most prosperous early settlements by Europeans, except Auckland, were fostered by the company or its offshoots.

Although the government had made it at least difficult and probably illegal to lease land directly from Maori, the law was largely ignored outside the formal settlements. Almost all the Wairarapa pioneers' capital went into sheep; initially, the land was the least of their financial problems.

Following pages: Merino sheep near Alexandra, Central Otago. Merinos from Australia were virtually the only sheep brought into New Zealand in the first years of European settlement, mainly because they had been hugely successful on the drier land across the Tasman, where, during the wool boom of the late eighteenth and early nineteenth centuries, the fine fleece of the merino was the most prized for export. New Zealand's wetter land meant foot problems, and when refrigerated shipping arrived, the value of the meat carcass became at least as important as its fleece.

Frederick Weld, one of the vigorous young men with capital who invaded the Wairarapa to set up the first sheep stations in the early 1840s. He later moved to Marlborough before making politics his profession. He became a New Zealand premier and then a colonial administrator as Britain's Governor of Western Australia, Tasmania and the Straits Settlements (Singapore and some Malaysian provinces).

According to R. D. Hill in *Pastoralism in the Wairarapa*, there were 'some seven hundred and eighty members of the Ngati Kuhungunu tribe in the Wairarapa when the squatters arrived'. They were pleased to see Pakeha because they had been decimated by invading tribes and knew the presence of the settlers would afford them some protection.

The valley floor and low hills were covered with bush and scrub, and grass and fern, with some long stretches of swampland especially round the lakes. It was the kind of flat, open land – comparatively rare in New Zealand and particularly so in the North Island – that suited large flocks of sheep. The trick was to get them into what was then a valley almost sealed off by mountains and the sea. The southern end was subject to the same southerly king-hit that afflicted Wellington but the ranges to the west kept all but the most southerly part fairly dry.

The first white men to peep into the valley were peripatetic Austrian naturalist Ernst Dieffenbach and the surveyor-painter Charles Heaphy who 'had a fine view of Palliser Bay and the Wairarapa Lakes' from a southern high point in the Rimutakas while on an 1839 expedition in search of the then rare and now extinct huia bird. Apparently the first European to wind his way around the south coast, over the Mukamuka rocks and into Palliser Bay, was William Deans early in 1840. He pondered the prospect of settling in the Wairarapa Valley, but then decided to go south. Later, he and his brother John settled on the then lone and level land of what became Canterbury, and set up the famous Deans Farm at Riccarton. The intrepid Ensign Best went hunting in the south of the Wairarapa Valley later that same year.

Two expeditions, in 1841 and 1842, really gave the Wairarapa a going-over for the first time by Europeans. First, New Zealand Company surveyor Robert Stokes walked in with a companion, J. W. Child, and two Maori guides. Another two New Zealand Company surveyors, Charles Kettle and Charles Brees, were next; and after an arduous month, which they probably wouldn't have survived but for food and guidance from local Maori, they walked out into the Hutt Valley over a route similar to that of today's Rimutaka highway. Brees reported of this pristine pastureland that it had 'an abundance of prairie grass and fernland for pastoral purposes; a sufficiency of timber for building and fuel', with 'quite sufficient land for arable purposes to suit settlers'.

In 1843 the group of would-be squatters did a deal with the Maori inhabitants. Frederick Weld, Charles Clifford, William Vavasour and Henry Petre (all from the same prominent English Catholic family) leased about one square mile of grass and scrubland and a huge area of hill country on the drier eastern side of the valley for £12 a year.

Others were involved with separate deals, notably Charles Bidwill and William Swainson Junior. These were the first to use the rough coastal route, carrying the sheep over the Mukamuka rocks one at a time – 350 of them – mostly in-lamb merino ewes. It was what Bidwill called a 'work of time' but one which 'was accompanied with less trouble than we expected and without any loss'. With autumn closing into winter, Bidwill and Swainson had passed the Weld party with their 600 merinos on their way round the wind-lashed, sea-thrashed southern beaches. The second group drove their sheep over the Mukamukas and lost some into the sea as the weather worsened. (The Mukamuka rocks were lifted high above the shoreline by the 1856 earthquake.) Thus the first two sheep stations in New Zealand were on grazing leaseholds held by Weld, Clifford and Petre at Wharekaka on the shores of Palliser Bay, and Bidwill at Pihautea.

A combination of the squatters' inexperience as pastoralists and the marginal adaptability of the merinos to the heavy, lush land made the start a tortured one. The Maori landlords moved closer and with their help and the hunting of game birds the newcomers were fed well enough through the first year until they could get their own gardens in shape. Then experienced Scottish and northern-England shepherds came to the rescue, first by moving the sheep off the lower land where merinos proved susceptible to foot problems. After seven years as a pastoralist, Weld was knowledgeable enough to write a bestselling pamphlet, *Hints to Intending Sheep Farmers in New Zealand*. It ran to at least three editions. Scab remained a problem, though, and labour was not easily available. The valley was especially good land for cattle, and before long some of the increasing number of settlers were complaining that too many of them were involved in the cattle trade for it to be profitable. Wool prices were climbing all the time, however, and profits were solid.

Within two years, twelve stations spread across the Wairarapa. By 1845 Weld, Clifford and Petre had 900 ewes and were wintering sheep for newcomers at Wharekaka Station. Petre sold his share to his partners and Weld began to find life hard in the isolation of the Wairarapa. He and Clifford decided to expand their business by buying a station in the Wairau Valley in Marlborough. Vavasour also went south. Weld and Clifford bought land at Flaxbourne (where the town of Seddon is now) and Weld wrote, 'I shall, with my yacht, be much nearer the town [of Wellington] than I am now. I shall have no rivers to ford, sometimes breast-high, no rocks to climb at high tide on the beach, or to sleep out in the rain all night. Nor shall I have any anchoring off lee shores in open boats, or swamping in the surf, of which I have had enough to last me for years.'

Wharekaka Station, on the shores of Palliser Bay, was established by Frederick Weld and his cousins, Charles Clifford and Henry Petre. By 1845, two years after they arrived in the Wairarapa, the station was carrying 900 ewes and wintering sheep for newcomers. Petre sold out but Weld and Clifford remained partners even after they moved to Marlborough a few years later. This pencil and watercolour of Wharekaka was done by James Coutts Crawford in 1846.

One of the reasons many of the pioneering sheep farmers in the Wairarapa moved on to Marlborough or Hawke's Bay after a few years was because of the shortage of workers. Shearing the large flocks on the big stations was labour-intensive and an arduous way to earn a living with hand clippers. Because it was seasonal, station owners were reliant on itinerant workers.

Sheep numbers built up with amazing speed on the native tussockland in the Wairarapa, Marlborough, Canterbury and Hawke's Bay. In the late 1840s Daniel Riddiford leased 30,000 acres on the east coast of the Wairarapa, which became Te Awaiti Station. His son, Edward, added more land and built up a huge flock of sheep, a part of which is pictured here at mustering. Over the following decades, the Riddifords became one of the wealthiest and most influential families in the region.

The first track was cut across the Rimutakas in 1847, and a road was completed for wheeled traffic in 1859. Until then settlers and their wool travelled to Wellington by sea.

By the end of the decade, Weld and Clifford had 11,000 sheep at Flaxbourne, and by the mid-1850s had more or less abandoned Wharekaka. Weld had a thirty-six-foot boat, rigged like a cutter, which took them across Cook Strait to Wellington and back.

Francis Dillon Bell, a Wakefield relative, was sent into the Wairarapa Valley in 1846 to buy 250,000 acres for a prospective Church of England settlement, presumably expecting incumbent pastoralists to pick up their sheep and disappear before the Lord's shepherds. Dillon Bell had written support from Governor Grey but made no progress with the Maori owners and got short shrift from the squatters. God and Grey willing, the Anglicans decided to go instead to Canterbury once the Free Church of Scotland settlers had rejected it in favour of Otago.

The Wairarapa remained under lease and it wasn't until the mid-1850s that the squatters managed to get title to their properties. Meanwhile, the wool grew fast and the profits faster. By 1850 many of the original pastoralists had quadrupled their investment. Weld, by 1852, had netted £4000, a heady sum in those days. Pastoralists stretched their enterprise into Hawke's Bay, which was almost entirely stocked from the Wairarapa, as well as Marlborough, Nelson and Canterbury. The squatocracy had arrived, educated men who had capital and the kind of confidence that Victorian Englishmen wore as their birthright. The world was their sphere. They were born to affluence and power. They called themselves 'colonists' and the other settlers 'emigrants'.

In 1855 sheep numbers in Marlborough and Hawke's Bay had edged past the Wairarapa. By 1861 Wellington (mainly the Wairarapa) had 250,000 sheep, fewer than the 312,000 in Hawke's Bay and 368,000 in Marlborough. Further south, the Canterbury flock had more than trebled in six years to 900,000 and in the same six-year period sheep numbers in Otago had, astonishingly, grown from 75,000 to 617,000. From then on the South Island was always easily the predominant producer of wool and sheep meat. The returns were spectacular through the 1860s. Wool exports earned £444,000 in 1860, £1,704,000 ten years later, £3,398,000 in 1875 but down to £3,169,000 in 1880 as prices began slumping. They did not pick up again for more than a decade. Sheep farmers' price woes were compounded by the spread of scab and the waves of rabbits destroying pastures.

Even more spectacular in the 1860s and 1870s was the increase in the value of grain exports from £18,000 in 1860 to nearly a million pounds twenty years later. The true wheat bonanza was the five years between 1875 and 1880 when the returns from grain (mostly wheat) quadrupled from £240,000 to £921,000. This explosion of grain exports was made possible by increasing farm mechanisation and the creeping tentacles of the railway system. It was also mainly a South Island enterprise.

The second wave of arrivals who set up the towns of the Wairarapa during the 1850s were the so-called 'emigrants', solid working-class types who followed a deal clinched by the government of the day to buy 150,000 acres of prime Wairarapa flat land for £2000 from local Maori. Towns were ruled up on maps by surveyors and colonising dreamers and the settlers moved in. The towns seldom matched the dreams of the planners but they became solid communities serving the farmers.

In 1853 the Wairarapa Small Farms Association was set up in Wellington by a group of men, led by Joseph Masters, in a bid to make sure land in the Wairarapa that they knew was about to be bought for settlement by the government from Maori was not given over in large areas to sheep farmers. As a result, two townships were drawn up and settled, the first named Greytown, after Sir George Grey, and the second Masterton, after Joseph Masters. This became typical of the two types of farming that then developed – the sheep farming of the large-scale graziers and the subsistence farming of small farmers in and on the fringes of settlements.

Local histories have described how hard it was for settler families in the very early days of the Wellington-Wairarapa region. A passage in *Brave Days*, written for the New Zealand Centenary in 1940 by members of the Women's Division of the New Zealand Farmers' Union (predecessor of Federated Farmers), says:

> In those early days journeys over the Rimutakas were memorable undertakings. Of the women, who with young children accompanied their menfolk on these treks, too much praise cannot be given. They were of the true pioneer breed, alike undaunted by the rough roads, flooded rivers, storms, privations and dangers of all kinds. In spite

The Wairarapa Small Farms Association was set up in Wellington in 1853 to provide for land-hungry Wellingtonians who did not have the kind of capital required to establish sheep runs. Charles Rooking Carter became the influential secretary of the association in 1853 and had a major impact on the destiny of the movement. The town of Carterton was named in his honour.

Eventually Hawke's Bay, Poverty Bay, Marlborough, Canterbury and Otago each carried many more sheep than the Wairarapa. Although the Wairarapa sheepmen launched the great age of the sheep stations many of them opted for other regions, and a landed gentry quickly blossomed in Hawke's Bay.

of their own trials, which in the nature of things must have needed more endurance than those of the men, they were always ready to encourage and inspire. It is therefore a strange thing that, in almost all the published accounts of those days, given by descendants of these settlers, little tribute falls to the lot of these noble women. They were there; they never failed their menfolk, and it all seems to have been taken for granted. Here and there we get a peep at them through the narrative, and these small glimpses reveal their initiative and steadfast courage.

Stories told of women having babies unattended in storm-isolated farmhouses, about families waiting for days to cross rain-enraged rivers, about people facing Maori bands lashing back at land-grabbers. The seclusion was very real at first and it was the ability of these women and their families to endure it and to produce wool by the tonne that put the mewling New Zealand economy firmly on the sheep's back. Wool started to earn New Zealand a stronger living in the world, the first renewable farm product to do so on any scale.

In the long term the tussock-tall grasslands of Canterbury and the valleys and hill country of Hawke's Bay, Marlborough and Otago carried many more sheep, but Wairarapa is where it all began. When Bidwill and Swainson carried their first flock of sheep over the Mukamuka rocks and drove them along the Pencarrow-Palliser Bay foreshore, they launched the age of the great sheep stations. Pastoral development rushed ahead in the south because its artificially planted settlements needed an economic base on which to survive and grow and because of the vast areas of tussockland surrounding them. The sprawling stations were to remain key economic units – despite a slump in wool prices that lasted from the late 1870s and through the 1880s that brought about a long depression – until the Liberal Party's Minister of Lands, John McKenzie, set about cutting them up in the 1890s.

The Small Farms movement

The first New Zealand Company settlers in Wellington, Nelson, New Plymouth and then Wanganui (a spillover from Wellington when the land shortage became critical) found the land they had been promised on the other side of the world was largely unavailable. The twenty million acres the company claimed to have bought were cut back by the government to less than 300,000 after the Treaty of Waitangi was signed.

In Britain, land ownership bestowed status and was insurance against pauperism. A. S. Thomson in his 1859 *The Story of New Zealand* wrote that settlers who had not got possession of lands as they expected were 'frittering away their lives in idle pastimes, while torpor and drinking had taken possession of a few'. He said little land was under cultivation and that farming implements were rusting for want of use.

Joseph Masters.

But among these wasters were many with access to very limited capital who seriously wanted enough land for subsistence farming. While many Wellington settlers grumbled and succumbed to torpor, Joseph Masters decided to do something about it. He was an industrious Englishman, born in 1802, who arrived in Wellington in 1842 and set up business as a cooper on the Lambton Quay beach. He and his son-in-law also began business as storekeepers and ginger beer manufacturers. His wife and daughter made straw bonnets. He was particularly scathing about Wellington's first settlers who were frustrated by lack of land:

> [They] were, with very few exceptions, very ignorant people, very few were practical farmers. Many of them were drunkards, and some worse; a number from the parishes and workhouses around London, together with a good sprinkling of Glasgow weavers. There were some good labourers from the provinces.

Masters had an adventurous background. After serving his apprenticeship as a cooper he became one of Sir Robert Peel's 'Peelers', a London policeman. He also worked as a soldier and a jailer. He spent ten years in Tasmania before coming to New Zealand with his family. A central figure among those working-class settlers in Wellington who wanted to buy land, he began writing letters to newspapers in the early 1850s advocating small farm settlements, especially in the Wairarapa. Governor George Grey was keen on the idea and the Superintendent of Wellington Province, Isaac Featherston, was a supporter.

Masters helped form the Small Farms Association, which sent him to visit the Wairarapa to choose suitable land and to persuade local Maori to sell. He claimed later that he lured Maori interest with the promise of shops like those in Wellington in new Wairarapa towns. Grey, not long before he left New Zealand for a new job in South Africa, arranged for 25,000 acres of land in the Wairarapa to be made available for sale. A ballot was held for lots of one town acre and forty acres for farming on the outskirts. Masters also organised finance. As a result, in 1854 the land on which Greytown (after Sir George) and Masterton (after Masters) were built was bought and settled. Charles Rooking Carter, a more recent immigrant also involved in the association, proved a steadier hand than the quarrelsome Masters and became the leading figure among the string of towns that quickly included Carterton and Featherston, established with the backing of the provincial government.

The land on which Masterton was built was originally claimed by a squatter, William Donald, who arrived on the heels of pioneering pastoralists Frederick Weld and Charles Clifford in the early 1840s, and whose descendants still farm sheep in the Wairarapa. William Donald's son, Donald Donald, invented a wool press which was essential equipment in many New Zealand and Australian woolsheds for 150 years. The manufacturing firm of Donald and Sons, formed in 1900, manufactured the presses and other Donald Donald inventions such as wire-strainers, saw benches and tractor-powered post-drivers.

The population of the towns doubled before the end of the 1850s, and Masterton was and still is the main town. The farmers who did best and survived were those who aggregated land by buying from those who gave up. Masters became unpopular not only because he bought up much land but also because he ran sheep on it – contrary to the spirit of the small farms. The Wairarapa remained predominantly sheep country, although the small farmers sold dairy products, grain and fruit and vegetables on the Wellington market. In the beginning, many of the subsistence farmers supplemented their income with outside work, often on road-making and maintenance, or on railway maintenance, and sometimes on the sheep stations.

In modern times, the Wairarapa has again attracted small farmers – of a very different sort from the people who followed Masters and Carter into the then inaccessible hinterland of Wellington. As the road over the Rimutakas was modernised and after a railway tunnel was cut through from the top of the Hutt Valley in 1955, many city-dwellers have moved out to distance themselves from the hustle and bustle as commuters, or to retire, or in search of an alternative lifestyle, or as agricultural entrepreneurs to live off the land in new, diverse and spirited ways.

CHAPTER FIVE

SHEEP IN THE MIDDLE ISLAND

Parts for Lister "No. 3" Shear

More than any other farming practice – even dog-trialling and wood-chopping – the competitive shearing of sheep fascinated townies and rural people alike in post-colonial New Zealand and Australia. The shearing culture, which was celebrated in poems and songs, grew from the arduous task of taking the wool off the sheep once or twice a year. The best, those who could do two hundred or more in a nine-hour day, were called 'gun shearers'. The support group included those who handled the shorn fleece and those who pressed the fleece ready for a compact bale. This is an historic shearing shed on Lagoon Hills Station, Wairarapa.

THE SECOND HALF of the nineteenth century belonged to the South Island – or Middle Island as it was called early on, investing Stewart Island with a geographical significance it couldn't sustain – and this was because of wool, wheat and gold.

The Free Kirk Scots had first choice of a settlement site south of Nelson. Lyttelton, Port Cooper then, was the front runner during the early planning. The Reverend Thomas Burns, a nephew of Robert and a stern family backlash to the philandering poet, wrote to the proposed expedition leader, William Cargill, in 1844, more than three years before they set out: 'I conclude that Port Cooper is our place of destination. I should be sorry that any change were made from that locality.'

Well, an irascible Quaker surveyor called Frederick Tuckett, entrusted with the on-the-spot choice of a site, changed their minds in favour of Port Chalmers. At first Burns suspected some dark conspiracy by Governor FitzRoy and the Church of England, but ultimately was persuaded the choice of name was a good one, comforted by advice that wood and water were not plentiful on the Canterbury Plains. And so Murihiku was carved up – between the Scots in Dunedin (from 1848) and the English in Canterbury (from 1850). The New Zealand-born politician and writer William Pember Reeves wrote of the two groups in *The Long White Cloud*:

When the stiff-backed Free-Churchmen who were to colonise Otago, gathered on board the emigrant ship which was to take them across the seas, they opened their psalm-books. Their minister, like Burns's cottar [farm labourer], 'waled a portion wi' judicious care', and the Puritans, slowly chanting on, rolled out the appeal to the God of Bethel:

God of our fathers, be the God
Of their Succeeding race!

Such men and women might not be amusing fellow-passengers on a four months' sea voyage – and indeed there is reason to believe they were not – but settlers made of such stuff were not likely to fail in the hard fight with nature at the far end of the earth; and they did not fail.

The Canterbury Pilgrims, on the other hand, bade farewell to old England by dancing at a ball. In their new home they did not renounce their love of dancing, though their ladies had sometimes to be driven in a bullock-dray to the door of the ballroom, and stories are told of young gentlemen, enthusiastic waltzers, riding on horseback to the happy scene clad in evening dress and with coat-tails carefully pinned up. But the Canterbury folk did not, on the whole, make worse settlers for not taking themselves quite so seriously as some of their neighbours. The English gentleman has a fund of cheery adaptiveness which often carries him through colonial life abreast of graver competitors.

The Canterbury settlers were led by educated men whose governance was well organised. With few fences over large tracts of land, a brand book was essential to identify flocks and Canterbury had one as early as 1861. Pictured is a page showing Samuel Butler's Mesopotamia brand, although misspelt.

For all Reeves' recommendation, some of the English settlers of the day did indeed take themselves with chosen-race, right-hand-of-God seriousness. An anonymous Indian Army officer who visited New Zealand in the very early 1850s and hurried back to India to hand in his commission and return to settle in Canterbury, wrote in a magazine article:

> It was a scheme to carry out complete, to an uninhabited land, a section of English society; to open the colonies to gentlemen; to transport knowledge and learning, the arts and the sciences; to prove to the world that the energies cramped by poverty, and the crowd at home, may have scope, by emigrating, without losing all that makes life valuable; and to build up a society on that foundation which experience had taught them to be best – the Protestant religion. It was a noble scheme, and Providence, as if to second such an effort, threw into their hands a tract of land perhaps better adapted than any other for the object they had in view . . .

The Canterbury Association had an even greater predilection for orderly and close-knit settlement of their province than the New Zealand Company had in other areas, and it had an even more rigid formula for the future. It wanted to establish a microcosmic England with a landed gentry, a middle class and labourers, all prayed over and preyed upon by the Church of England, but without the poverty and ignorance that pervaded the lower levels of British society at the time. It was sheer Utopianism to plan a social and economic structure and expect to avoid the by-products of the society it was exactly modelled on. But as in Wellington, reality intervened, and the plan lived on as only a plan, mostly as an illusion. The association felt able initially to charge £13 an acre for land in the new settlement – one-third to be granted to education and the Anglican Church, one-third to be spent on immigration, a sixth on administration and public works, and the remaining sixth to go to the New Zealand Company for the land – a 500 per cent mark-up.

Governor Grey did not like the price much, especially the slice allocated to the church, and the settlers liked it so little they bought only £50,000 worth in the first year, 1851, instead of the expected £1 million. What actually happened was decided, as it usually is, by a combination of the ingenuity of man and the economic opportunities he discerns. Settlers were not to be confined to town by a philosophy of settlement. Squatting commenced on a large scale. The sheepmen moved out quickly from Christchurch and soon great flocks were spreading over the plain. Clifford and Weld had taken sheep into the open country of Marlborough as early as 1846 and were joined in the range-grazing of sheep by a number of leaders of the Nelson community. They became a major source of sheep for the Canterbury runholders in the following decade, although imports from Australia continued for a number of years.

In the 1850s the price was still relatively high for wool, and demand for increased quantities, although it wavered on occasions, remained basically strong; runholding held up economically until well into the 1860s. New Zealand's all-time best-known sheep farmer, Samuel Butler, arrived in January 1860 and when he left in the middle of 1864 he had doubled his investment in his Canterbury holding at Mesopotamia. Three or four years later he might not have been so lucky. Some sheep farmers were ruined by the big snow of 1867. The reasons for wool's buoyancy were complex, but certainly involved was the discovery of gold in Australia, which drew hundreds of men off the land in the fifties in search of a fast fortune. Also a factor was the accelerated movement of Englishmen and Scotsmen off the land and into the cities in Britain as the Industrial Revolution gained pace and outstripped the agricultural revolution.

The Wairarapa pastoralists and their shepherds had been predominantly Englishmen, with a few Scots. But from the first years of the Canterbury settlement a new influence arrived: Australian sheepmen. L. G. D. Acland wrote in *Early Canterbury Runs*:

> There had been a very bad drought in Australia in 1850 which disgusted many of the squatters there with the country. Some of them sold or abandoned their runs and came

The most famous sheep farmer New Zealand has ever had was Samuel Butler, who settled on his station, Mesopotamia, in 1860 and had doubled his money when he left four years later. His timing was perfect. Later in the decade, some runholders were ruined by big snows and diseases. Butler used the mountains behind Mesopotamia as the setting for his famous satirical novels, *Erewhon* and *Erewhon Revisited*.

to New Zealand early in 1851 to try their luck again with what money and stock they had left. They not only brought money and stock but experience, which was most valuable to the young colony. They had unlimited faith in the squatting system, and a great contempt for the Canterbury Pilgrims' desire for freehold agricultural farms. The Australians were nicknamed 'Prophets' or 'Shagroons' and it was they who nicknamed the association's settlers the 'Canterbury Pilgrims'.

'Shagroons' was the name that stuck. These tough and taciturn men, hats hung low on bobbing heads, almost appendages of their stringy Australian horses, drove their sheep past the less assured Canterbury Pilgrims and led the thrust onto the plains beyond. Their machismo was not an affectation. They had endured the dogged dust and distances of Australia and indomitably mustered their sheep for the hard trip to New Zealand when their holdings in that other country could not be tamed. They knew all there was then to know about sheep, especially about merinos, the predominant breed in both Australia and New

Zealand at that time, and they were notoriously impatient of any restraints upon grazing the unpopulated plains. William Pember Reeves wrote, 'Coming to Canterbury, Otago and Nelson, they [Australian sheepmen] taught the new settlers to look to wool and meat, rather than to oats and wheat, for profit and progress. The Australian coo-ee, the Australian buck-jumping horse, the Australian stock-whip and wide-awake hat came into New Zealand pastoral life, together with much cunning in dodging land-laws, and a sovereign contempt for small areas.'

A number of Aborigine and part-Aborigine stockmen came out with the Australian squatters and worked for them and other early South Island runholders. The Otago historian Herries Beattie claimed that as a young man he often heard sheep called 'jumbucks', as they were known in native Australian dialect.

By temperament and background many of the shagroons were more akin to whalers and sealers or the bushmen of the North Island than to the genteel upper-middle-class Canterbury Association settlers. There was a mutual contempt too between the agricultural men and the pastoralists, as there had been in the Wairarapa. For the Australians, sheep farming was the manly way to live on the land and the good news they soon discovered was that, because of the fast-growing grass in a more benign climate, merino fleeces averaged four pounds a head in Canterbury compared with around two and a half pounds in New South Wales. Years later Herries Beattie wrote, 'In pastoralism we copied Australian ways rather than those of other lands.' But he noted that the term 'squatter' was not used 'except jocularly' in the South Island and that the rest of the Australian pastoral patois disappeared quickly.

Since the 1840s an argument raged around whether the leasing of land was illegal, but in late 1851 the Canterbury Association Resident John Godley changed the rules and allowed unused land to be taken up by runholders in lots of 5000 to 50,000 acres, according to locality. That set off a scramble for land, the plot of so many Westerns, told matter-of-factly by Butler in his *A First Year in Canterbury Settlement*. The procedure as Butler and others have described it was to ride out and stake a claim, then apply to the land office and if your claim was first the run would be yours. The runholder then had to stock it within six months, one sheep to every twenty acres or a head of cattle to every 120 acres, and he had to pay a farthing an acre rent for the first two years, halfpenny an acre for the next two, and three farthings for the fifth and all subsequent years.

Acland added:

> If he did not fulfil these conditions, he forfeited the run. No term was stated to his licence, but he or anyone else could buy the freehold of all or any part of the run at any time, except that if he built a hut or made a fence or put any other improvement on the run, the improvement gave him the pre-emptive right over so many acres adjoining it. Anyone could challenge his pre-emptive rights at any time, and if the owner didn't buy the land challenged within a month the other man could, and then the runholder got nothing for his improvements.

Opposite: The ruins of the Teviot Station woolshed, built in 1880 and destroyed by fire in 1924. The capacity of the shed was 8000 ewes with forty shearing stands. It was said to be the largest woolshed in the southern hemisphere and the largest stone farm building in New Zealand.

Teviot Station was taken up in 1857 by Captain John Cargill (son of William Cargill). In those days it was 64,000 acres and carried 42,000 sheep.

Below: John Deans had a look at the Wairarapa before deciding to farm on the Canterbury Plains, at Riccarton. He and his brother, William, built the first houses on the Plains (below left) and were well established by the time the sheepmen arrived with their big herds a decade later. Their descendants are still prominent in Canterbury.

Gaining and retaining land through the period of squatting in Canterbury was difficult. It called for an understanding of complex legislation and a deep and ruthless cunning. It is best described at length by R. M. Burdon in his *High Country*.

Livestock had been farmed in the South Island since the 1840s, with small flocks of sheep, larger herds of cattle and also horses imported by whalers. The ex-whaler Johnny Jones had about 2000 sheep, 200 cattle and 100 horses at Waikouaiti in 1844, on a property called Matanaka. Banks Peninsula was settled in 1841, and the Deans Brothers were well settled with stock at Riccarton by 1843. From 1850, with the spread of the runholders, sheep and cattle grazed the grass and tussockland of Otago and Canterbury. By 1854 an estimated 100,000 sheep thrived in Canterbury and four years later more than half a million; and by 1858 a quarter of a million grazed Otago tussock. At that time, Canterbury was either deep swamp or heavy fern country along the coast, with large tracts of tall manuka scrub behind it, backed by heavy tussock further inland.

A number of stations were first stocked with cattle and some farmers continued to breed and fatten large herds. Following the discovery of gold in Central Otago, and later in Westland, New Zealand's population soared from 60,000 to 200,000 in the 1860s, creating a growing demand for meat; but until refrigeration became a commercial proposition in the late 1880s, cattle did not make the economic sense that sheep did. The heavy trampling of cattle and their ability to survive on more rank growth than sheep made them a sensible animal for grazing new country, but once the natural pastures were improved or English pastures patched in by burning and sowing, then sheep did better. The use of both animals for complementary grazing was generally not practised in those very early days. Acland wrote that by the 1860s there were about 250 stations in Canterbury, and that perhaps twenty of them, mostly along the coast, carried cattle.

Every station had house cows and, before tracks were flattened enough for the up-tempo pace of horses, bullocks were the main draught animals – muscling inexorably along in front of a sledge or dray. Training and working the bulls was an especially tough job and bullockies became notorious for their ear-blistering language. Bullocks were almost exclusively used for first-time ploughing of New Zealand land.

The Wairarapa pioneers, Clifford, Weld, Petre Vavasour and Bidwill, the first to bring sheep into New Zealand in substantial numbers, farmed only merinos. The breed originated in Spain, and came to this country direct from Australia after being influenced as a breed by farmers in a number of other countries, notably France, Britain and South Africa. The merino is tough, active, wiry, and nervous unless up in the hills with plenty of room to move, but susceptible to footrot on heavy ground. Its wool is compact and fine, its meat lean and stringy. It was not the ideal sheep for the spongy valley floor of the Wairarapa but up in the hills it thrived, as it did on the drier hills and plains of the east coast of the South Island, Central Otago and Hawke's Bay. Before refrigerated shipping began the export meat trade in the 1880s, there was an enormous surplus of mutton in this country. Merino carcasses were boiled down for tallow but that didn't matter; their sole reason for existence then was to grow wool. Merinos were adept at fossicking for food in the often rank tussock country and were less troubled in the South Island than in the North by footrot. The sheep were driven onto new land under the eye of a shepherd and, with luck, natural boundaries to the run made the task of keeping the flock at home easier. In some cases, dogs were tied on long leads to boundary posts to bark sheep back to their owner's property.

An early task for the settlers was to replace tents with whare, primitive at first, possibly wattle and daub or sod, thatched with raupo or tussock; then perhaps another whare fairly soon after, at a strategic place, for a shepherd. The next stage would be a cob homestead, followed ultimately by timber with iron, slate or shingle roofs. The runholder simply put his sheep on the run and directed most of his energies to keeping them there. He marked the sheep when necessary and sheared them when it was convenient. With ewes and rams often running free in hard country, mustering was difficult and the times of lambing and shearing were hard to control. Lambs dropped between May

Opposite: The *Illustrated New Zealand News* carried 'Sketches of a Sheep Station During the Wool Season' in 1883. The individual sketches are described as 'At the Wool Table', 'Old Jerry the Pet', 'Wool Pressing and Classing' (a bundle of wool is being tipped into the press), 'Branding and Cutting Horns', 'Drafting Sheep with Pets' (who act as decoys into the yards), 'On the Way to Town' and 'Shorn Sheep in Pen'. 'Pets', or bellwethers, were used frequently before dogs took over to muster flocks.

and August suffered heavy losses and every year some sheep would be mustered up with two years' fleece, having been missed in the muster the year before. The earliest runholders who pioneered large tracts of country in the South Island had a special range of problems: wild dogs attacking stock; rats, mosquitoes, fleas, sandflies and blowflies scourging the men and their supplies; the poisonous tutu plant (mainly near the coast), speargrass and the thorny matagouri; floods, swamps, bridgeless rivers, pathless hills and mountains; and, always, isolation. The loneliness was often long and intensive, as Herries Beattie tells in *Early Runholding in Otago*:

> Most accounts of pioneer life on the runs stress the loneliness. Women had a very lonely time of it. The first woman on the Waikaka Run went there in December, 1858, and lived in a 'warrie' [whare] at what was later the junction of three populous roads, and yet she was there twelve months before she saw another woman. The same experience befell another woman a dozen years later on an out-of-the-way run in the Hokonuis. Two travellers through Otago Central in 1859 saw only one woman. When she saw them in the distance she tore down on horseback to see if, by any chance, her husband was one of them. He had ridden away to get cattle four months earlier, and she had heard nothing of him. When asked if she was not afraid to live alone in such an outlandish locality, she nonchalantly replied, 'Oh, no! There's no one to hurt me!'

Scab was the first scourge of the sheepman. It came before the rabbit menace and proved easier to eradicate. It was imported originally with merinos from Australia, was known in Nelson in 1845, the Wairarapa in 1846 and at the Deans' farm, Riccarton, in 1848. It spread swiftly through Canterbury and Otago where stock on the drove mingled with sheep on unfenced runs. Scab was an infection under the skin that set the sheep to rubbing itself against anything it could find, thus helping to spread the contagious disease. The sheep would rub its wool off and gradually lose condition. Runholders took to the problem with tobacco and sulphur dips and by inducing the government to invoke stringent laws against those who did not attack the disease or those who spread it by moving infected animals. Scab was at its peak in the 1860s

By the time hundreds of thousands of sheep were grazing the Canterbury Plains, the demand for seasonal labour was heavy. Many were itinerants who moved down the country as the summer began, shearing as they moved south. Just how labour-intensive was the work is demonstrated by this drawing made in 1864 of shearers, fleecers, and woolpackers.

The flat land on the Canterbury Plains had all been taken up by the end of the 1850s and the sheepmen began to move into tougher country on the hills and in the elevated Mackenzie Country (pictured). Big snows on the flat land ruined many a runholder in the 1860s and 1870s but the intrepid men who moved sheep onto the hills and the mountain valleys learnt to cope with the severe winters and unseasonable snows.

and gave a fresh fillip to fencing, as galvanised wire became plentiful. The tough legislation worked. By 1892 the disease was officially declared eradicated.

An intransigent attitude to a fatal catarrhal disease of sheep, which was rife in Australia at the same time, kept it out of New Zealand and focused the attention of farmers in this country on the effectiveness of quarantine laws. As soon as the threat of a disease became known, farmers and the government moved swiftly to keep it out. A parliamentary select committee in 1861 met to take evidence on whether the importation of cattle should be prohibited from those countries where pleuro-pneumonia was known to exist. It recommended accordingly. Parliament, by the way, not only acted with alacrity to a threat in those days but dealt with crises expeditiously. The select committee met on 13 June 1861, heard evidence from a number of farmers, butchers and veterinarians, and reported to Parliament on schedule – eight days later.

The government took swift action again in 1866 to prohibit animal imports of any sort from countries known to have rinderpest, at that time raging in Panama. The legislation set up the framework within which emergency measures could also be taken to stop the spread of the disease within the country if it arrived. Thus early advantage was taken of New Zealand's isolation in the world to protect its herds and flocks from the fatal or debilitating effects of many of the worst animal diseases. Successive governments – aware that the economic stability of the country depended on the ready acceptance overseas of disease-free animals and animal products exported from this country – maintained an implacable quarantine regime.

By 1858 all the flat land in Canterbury Province had been taken up. It was still understocked but it was accounted for. Settlement by squatting was spreading swiftly westwards to the foothills of the Southern Alps and about the same time was fanning out from Dunedin into Otago and Southland where conditions were often even more rigorous and isolated than in Canterbury. About 138,000 acres of the country, mostly in the South Island, had been ploughed or burnt off and sown in pasture in 1858. Twenty-five years later, more than two million acres had been sown in pasture and average fleece weights had doubled over the same period.

The big squatters, leasing thousands of acres each, developed a kind of divine-right-to-land approach to politics in Canterbury. When Grey dropped the price of land to five shillings or ten shillings an acre, depending on locality, the idea was probably to give more men access to land, thus preventing too large aggregations. But the shrewd and sophisticated runholders fought back with spotting (buying the best spots on their huge leasehold tracts to make intrusion difficult) and gridironing (buying strips to make access difficult or in such a way as to leave less than the minimum twenty acres for an outsider to buy). And because they were accumulating capital faster than any other group anyway, the lower prices meant they could amass larger areas than ever. 'It is a strangely ironical fact,' wrote R. M. Burdon in *High Country*, 'that at the very time when he [Grey] was making stirring speeches about the evils of land monopoly, the greatest amount of land accumulation was going on.' Or as William Pember Reeves put it: 'In the earlier years the squatters were merry monarchs, reigning as supreme in the Provincial Councils as in the jockey clubs. They made very wide and excessively severe laws to safeguard their stock from infection, and other laws, by no means so wise, to safeguard their runs from selection, laws which undoubtedly hampered agricultural progress.'

The intelligence, wealth and arrogance of these men made land the prime political issue right through into the twentieth century and created the powerful farming lobby, dominant nationally until the 1960s. 'The second generation of large graziers rarely inherited either the charm or the education of their fathers,' wrote Alley and Hall, 'but at least they produced an increasing quantity of wool.'

Yet the offspring of the graziers may not have been quite as grey and undistinguished as this assessment implies. A number of them who had special talents or academic skills defected to England, which they still thought of as their home. This sense of identity with the British lasted long in New Zealand. The poor and rejected who formed substantial numbers of the immigrants to Australia and the Americas had less cause to remember their origins than the significantly better off, more respectable 'colonists' of New Zealand. It was part of the national state of mind and consequently part also of the national literature, as an early poem 'Emigrant' by Mary Colborne-Veel illustrates:

> Homely flowers set
> Where our farmsteads rise,
> Make an England yet
> Under sunny southern skies.
> Lilac scent is blown
> With wattle on the breeze;
> September bids the leaves grow broad
> On happy English trees;
> And apple-orchards smile again
> In sweet, familiar show –
> But in my heart is mourning
> For the scenes of long ago.

By centennial year, 1940, most New Zealanders had been born here, but so many church, political and academic leaders were from Britain that the British thing was still pervasive. Writers such as Robin Hyde and A. R. D. Fairburn were indelibly, if a little self-consciously, New Zealanders first and their literary voices reflected that; but they also belonged to a kind of subculture. The University of Auckland College published a book in 1940 called *1840 and After*, subtitled 'Essays Written on the Occasion of the New Zealand Centenary'. It may seem extraordinary now, but most of the authors, all from the staff of the University of Auckland, ignored this country's existence. In his foreword, the college president, Mr W. H. Cocker, said, 'English culture of the early and middle nineteenth century has a special significance for New Zealand and her history. The authors of this book . . . have therefore turned their telescopes upon Victorian England. What kind of civilisation was it that was brought to this country? . . . The reader of this book will have a reasonably complete picture of what was probably the most significant period in New Zealand's past.'

Especially in Canterbury, but also later in Hawke's Bay, runholders who achieved success and affluence tried to emulate the kind of life lived by the landed gentry of England. They achieved a sort of elegance, the more impressive for its remoteness from civilisation. As Lady Barker put it in *Station Life in New Zealand*:

> The older I grow the more convinced I am that contrast is everything in this world; and nothing I can write can give you any idea of the delightful change from the bleak country we had been slowly travelling through in

The men who gambled on wool by staking out huge areas and running sheep over them were often adventurers who enjoyed taking risks. Most were upper-middle-class Englishmen or Scots but some were self-made and they were often the most colourful. One of the most famous was William Robinson, son of a tenant farmer in Lancashire, who made quick money in South Australia as a stock dealer and overlander. He bought a huge run there and became a leading pastoralist. He sold up in 1856 and soon had control over 84,000 acres around Cheviot. He lived in Nelson, and later in Christchurch, and became a runholders' advocate in the Legislative Council. He was hugely successful and was dubbed 'Ready Money' Robinson because he made money quickly and lavishly bought land, horses and other stock.

pouring rain, to the warmth and brightness of this charming house [near Amberley] . . . The inside of the house is as charming as the outside, and the perfection of comfort; but I am perpetually wondering how all the furniture – especially the fragile part of it – got here.

The good woman describes a picnic that is a sedulous transference to New Zealand of an English manor-house ritual. But elsewhere she refers with kindly condescension to the poor quality of servants in Christchurch, the inadequate music and unsophisticated behaviour at balls.

Many runholders aspired to a high life because of their background, but they had paid their dues, as Alley and Hall said:

The men who took up the early sheep stations were often people of education, belonging to families of English country gentry, or at least to those connected with the services and the professions. In the early years these men did not hesitate to put up with extremely crude living conditions – the slab or cob hut thirty miles from a neighbour. Men reading Latin or Greek for pleasure lived the hard, self-sufficient life of the back country with gusto, still hoping great things for the future.

A kind of egalitarianism nevertheless developed, not as a philosophy but as a fact of the existence of the new colony, a fact commented on by many early visitors. Often the motive for emigration, even among the middle and upper-middle classes, was a social incompatibility at home in Britain – some rejection or failure – and no matter how much the actual decision to come out was made by the emigrant himself, it was still a kind of exile. A figure from the folklore of New Zealand was the stereotype mystery man of good education, working as a station hand, provoking speculation that he was of noble descent, or at least from the upper class, because he spoke so well. He would go to town periodically and drink away his earnings, or his remittance from 'Home', then come back to the station and work hard until next time.

'Wonderful tales are told of cultivated men in the wilderness,' wrote Reeves, 'Oxonians disguised as station-cooks, who quoted Virgil over their dishwashing or asked your opinion on a tough passage of Thucydides whilst baking a batch of bread. Most working settlers, as a matter of fact, did well enough if they kept up a running acquaintance with English literature; and station-cooks, as a race, were ever greater at grog than at Greek.'

When transport and communication were slow and arduous – because of the intervention of so much time – and inevitably imprecise, it was easy for a man to dramatise himself, if only by innuendo or by what he didn't say. Facts were dilatory travellers to the advantage of romanticisers, dreamers, frauds and swindlers. According to *The Age of the Great Sheep Runs*, an essay by P. R. Stephens:

There was little that was patriarchal in the relationship of the runholder to his employees and on many properties a somewhat suspicious, harsh attitude, probably imported from Australia, was the rule. Accommodation was poor, and the food often little more than unlimited supplies of tea, bread and mutton; for the whole system was based on the existence of a large number of itinerant casual workers. Shearers were in a better position to assert their rights and some fairly bitter disputes took place. The Amuri Sheep Farmers' Association during the late 1880s even asked the government to import shearers from Africa.

Even so, a code of hospitality often did prevail, with food and rough shelter made available to itinerants by some farmers and their families. But life on the road as a casual labourer or shearer was spartan – men mostly without women, and little social life outside the pub, and with alcohol used freely to numb despair and blunt aspirations. Hardly surprisingly, alcohol became an enormous problem affecting the health and well-being of thousands. Accounts of early life in New Zealand often refer to alcohol as the major health problem of the time.

On the swag

For many decades from the last quarter of the nineteenth century, itinerant farm workers followed the seasons in Australia and New Zealand. They were swaggers 'on the swag'. The swag contained a small tent, a sheet of oilcloth to cover the damp ground, a billy for tea, and all the rest of their worldly possessions.

New Zealand politician and writer John A. Lee was on the swag as a very young man and wrote of the life in a rich and funny book, *Shining with the Shiner*:

> They came regularly from Australia for the shearing, the harvesting, in large numbers . . . They walked to work. If there was no work and no dole, time on the road did not much matter. It was a cheap way to get by. A fortnight would take a man across most of the South Island. They erected calico tents in the trees near the fresh water and were ready to start when the season and the boss willed . . . And the concertinas might come out around the fire and voices might sing 'Alice Ben Bolt' or 'The Wild Colonial Boy'.

Lee was one of many storytellers of the time, encouraged as they were by sitting around the fire in the evenings with no entertainment but each other. It was a hard life between drinks. Runholders hired special workers to set the pace and those who couldn't match it were soon sacked. Among the swaggers were remittance men, members of wealthy families who were sent to the colonies because they had committed a crime or were generally troublesome and a threat to the family reputation. They would periodically line up at the post offices and banks to receive money remitted from families back 'Home'. Many were alcoholics who would work hard to supplement their remittance and when the job was done almost all would go on the booze. Every sheep station had its story of the well-spoken, educated permanent hand who would disappear to pick up his remittance, go on a bender and hope to still have a job when he returned. Lee wrote:

> And as the country grew settled and the telephone and good communications came and as the steady among the industrious, for all were industrious if some were improvident, settled into jobs . . . The men who remained on the road were the men who drank as hard as they worked, the tough and strong who were the cageless, the unadaptable, the artists who in a strange country had become labourers; and of this type there were many . . .
>
> In those days the gap between vagabondage and respectability was not immense, a bad season, a foreclosed mortgage. How many a prominent citizen of yesterday have I seen settle down, fill a pipe, and start: 'When I carried the swag . . .'

A swagger waits at the kitchen door to find out whether he can have a job, or a meal, or both.

Lee, on the road as a boy after escaping from borstal, was the most famous and the most published of the New Zealand itinerants and, like many of them, got off the road to find steadier work. The experience of living rough with hard-core alcoholics encouraged him to be a teetotaller.

An even larger tradition of the itinerant storytelling swagger spread through Australia. Many men travelled among the states and included New Zealand in days when their common remoteness held the two countries closer together than today.

One of the last of the swaggers was 'Russian Jack', still on the road in the Wairarapa until the beginning of the 1960s, by which time he was what we would today call simply 'homeless'. He died in 1968 aged ninety.

Cabinet minister and writer John A. Lee was the most famous of the former itinerant workers who made good in later life.

CHAPTER SIX

GRAIN AND INGENUITY

The leader of the Otago settlement, William Cargill, was alarmed at the prospect of a goldrush, which he and his sober, senior associates thought would attract ruffians to the town.

THE GOLDRUSHES to Central Otago, and later Westland, and sliding wool prices brought about a move to wheat-growing and a boom in other grain crops as well. The Free Church of Scotland settlers of Dunedin had been wary of immigration in the 1850s from what they saw as undesirable sources, and those sources included Catholic Irish and anyone from the waning goldfields of California and Victoria. But Central Otago had already attracted prospectors because the big, buckled landscape with fast-flowing rivers looked promising to the practised eye. After Tasmanian Gabriel Read found gold at Tuapeka, the quick prosperity it brought to a province in a slump soon persuaded the people of Otago that gold was no bad thing at all, whether it attracted ruffians or not.

The first report of gold in the province arrived in Dunedin in 1851 when the settlement was only three years old. Two men looking for grazing land claimed to have found some. The first firm discovery was in 1856 when the Surveyor-General, Charles Ligar, said he had found widespread traces in sand and gravel during a visit to Mataura. Glimpses of 'colour' were also seen in other parts of the province during the rest of the fifties. There were no big finds though – the land was being coquettish.

Dunedin's settlers heard Ligar's news with what Otago historian A. H. McLintock called 'ill-concealed alarm'. The Scots wanted to preserve their settlement from temptations within and shore it up against corruption of their values from without. When gold was not more than glittering talk, settlement leader William Cargill and his sober, upright cohorts could foresee the boots of 10,000 diggers trampling their cultural and religious purity underfoot. Yet when Read made his discovery half the sober population of Dunedin was off and running for Tuapeka.

Gold had an enormous impact on the farming of New Zealand. The population of Otago in 1861 was a shade over 12,000. Ten years later it was close to 60,000. More than £23 million had been pulled out of the ground, and the hinterland of Otago was opened up – in some cases left scarred beyond recognition. The influx of labourers and artisans into Dunedin liberated social and political thinking from the dominance of the conservative church men and the pastoralists whose power had grown through the 1850s. Gold also consolidated the economic superiority of the South Island over the North, already established by the sheepmen. It wasn't until the turn into the twentieth century that the North Island began to catch up.

The population of the whole country during the golden decade moved from around 60,000 to 250,000. With new national wealth from the export of gold (which peaked at £2,845,000 in 1866, more than wool earned that year) and tens of thousands more people, farmers could shift a little from growing wool for export to producing for the local market. Around the goldfields, agriculture had swiftly followed the population build-up with a sudden stretch of acreage under consumer crops. It wasn't just for food. Barley production soared as men without women – the miners who crowded in

– found solace in booze. The ratio of women on Central Otago goldfields was 1:100 men in 1861, although it grew to 18:100 in 1864, 30:100 in 1867 and 47:100 in 1871. (Over the whole of New Zealand in 1861 women numbered only sixty to a hundred men.) Seventeen breweries were operating in the province by 1871, producing more than forty per cent of the country's output of beer.

So gold and its direct effects, along with problems managing sheep, price fluctuations for wool, and the development of cultivating and harvesting machinery meant grain waved across the South Island countryside in the seventies, eighties and nineties. South Canterbury farmer and poet F. H. Wood wrote:

> The seas of wheat o'er which the wind-made wavelets
> Sweep on and break upon a phantom shore.

Canterbury and Otago were the provinces – and still are for that matter – most suited to the growing of grains. Farmers there had the climate, the soil, the flat acreage, the capital, the organising ability needed for large-scale cultivation, and access to the extra labour required for wheat. And they had the big thinkers, the risk-takers, too. One of the virtues of the early New Zealand farmer was that he had no traditions to tie his mind up. Many of the pioneers had little previous experience and what they had needed testing against a new environment in a new country.

Otago's Donald Reid, who left his lowly home in Perthshire at the age of fifteen, had twenty prime acres at Forbury, in Dunedin, at eighteen, 200 acres at twenty, and within twelve years of his arrival was cultivating 500 acres. With farm produce, general stores and fast transport he pulled more money out of the goldrush than all but a few of the diggers – and he even got a share of the actual gold. As his farm labourers prepared to hotfoot it for the goldfields, Reid counselled patience, formed a co-operative group and they all got their share from a rewarding claim worked on their behalf. Characteristically, he was rapt in the potential of machinery for cultivating and harvesting grain. Alley and Hall described the reaping machine Reid bought in 1861:

A farm worker sharpens his scythe before attacking a paddock of grain in the days before demand from population growth brought machinery.

It belonged to that strenuous old type from which the cut grain had to be raked off sideways. He soon replaced this with a back-delivery reaper. In 1862 he used his profits from gold, including a share in an actual claim, to buy machinery in quantity – a chaff-cutter, corn-crusher, a threshing machine and a steam engine. A more powerful steam threshing plant was already at work in the Taieri, but Reid preferred to be able to do his own work in his own time and to hire out his machinery to others.

From about 1865 he gradually changed over from cropping to sheep farming, his first highly profitable herd of cattle having suffered a good deal from stock disease. This flexibility is typical of the most alert farming minds in early New Zealand. Reid and others resumed the production of grain on a large scale in the seventies, when it developed into a substantial export industry. Meanwhile the

Threshing machines were available to wheat farmers surprisingly early. The one pictured is working at Temuka in 1861. It had been landed at Timaru by whaleboat after use in the Christchurch area. (This was the first photograph printed in South Canterbury.)

pastures he laid down in English grasses carried increasing flocks of sheep.

John Grigg and Duncan Cameron were two of many Canterbury men who backed their practical skills and foresight by moving into wheat when the time was right, and back into sheep again when refrigeration looked a prospect. Not that sheep farming suffered greatly during the wheat boom. Flock numbers kept up and fleece weights for each animal doubled during the last quarter of the nineteenth century. The families and estates of these agricultural entrepreneurs may now be institutions firmly grafted to the establishment, but in their day they were the highest fliers New Zealand has produced in business. Their imagination and courage were applied to big risks and big stakes. They won most of the time and New Zealand shared the prize. Rich and resourceful Grigg was always at the forefront of mechanical and business innovation in Canterbury and his large wool clip went to market in style, powered along a dirt road by a huge steam engine.

Like the Americans of their time, these Canterbury and Otago farmers had great faith in machinery and they had the support of imaginative blacksmiths, mechanics and engineers. Interest was so intense, a reaping match was held in Hagley Park, Christchurch, in 1864, to decide on the efficiency of six different types of imported and home-made machines. Each had to cut one and a half acres of a fairly even crop of oats. First prize went to a Burgess and Key (imported) hand-delivery reaper which beat the first McCormick reaper. (They were reapers and not reapers and binders.) The first reaper and binder was imported in 1877 by a Christchurch blacksmith called John Anderson and, according to an article in the Centennial Supplement of the *Christchurch Press* in 1940, 'These machines did away with labour and tales are told of barns being broken into and machines being broken up by labourers who feared they would soon be out of work.' But resistance from the Luddites was slight. In the same year, the first American seed drill arrived. Business was brisk. The records of one firm (Morrow, Bassett and Co) show that fifty McCormick harvesters were brought out in 1879 and 1880, and even though the price was high at £75, the supply was not enough.

Anderson set up as a blacksmith in 1850, and in 1857 began making farm implements, many of them to his own design. He sent his two eldest sons to Edinburgh in 1866 and they returned a few years later with degrees in civil and mechanical engineering. He was a man of great diligence and high imagination: during his blacksmith apprenticeship days in Edinburgh he worked from 6 a.m. to 6 p.m., but found time to gain a diploma from the city's School of Arts as well.

Wool prices were wavering at the beginning of the seventies; and during the 1860s farmers, especially in Canterbury, had their first serious climatic difficulties getting stock through the year in open grazing. After the benign fifties, two sheep-burying snows and a killer flood made the decade arduous and ruined some runholders. So ploughshares bit deep into the better land at the beginning of the seventies and the 'bonanza' wheat farming began, almost ninety per cent of it in the South Island. Not that the price was encouraging – it was actually down as the eighties began and as it went down the acreage went up. It was a time of economic depression. In this first period of wheat's 'high importance', wrote Hilgendorf in *Wheat in New Zealand*, prices averaged only 3s 4d (34c) a bushel, over a twenty-year period, 'yet wheat growing paid better than anything else. Cultivation had become mechanised with the introduction of the drill and the binder, and thus large wheat growing farms came into existence.' John Grigg's Longbeach property had 3000 acres under grain, 120 working horses, and twenty ploughs of two or three furrows working on one field. Duncan Cameron of Springfield Estate, near Methven, grew wheat on a comparable scale, and many others approached it.

Despite mechanisation, wheat-growing was still a labour-intensive business, even compared with wool-growing, which had its annual peaks of mustering, dipping and shearing. Shearing required additional, hard-working labour for a short period and it was the major organisational problem that a runholder built his year around. Alexander Bathgate wrote in *Colonial Experiences*:

> When this great business of the year is over, the squatter often allows himself a holiday,

The first reaper and binders became available in 1877 and, according to the *Christchurch Press*, attracted Luddites because they reduced the demand for labour. Barns were broken into and machines smashed over the following few years. The reaper and binder pictured was cutting oats on a farm near Levin in the first decade of the twentieth century.

Grain and Ingenuity

BEGINNING · of SECONDARY · INDUSTRIES · JOHN · ANDERSON'S · FORGE · CASHEL · St 1851

John Anderson was one of the pioneers of agricultural engineering in New Zealand. In his Christchurch forge, he designed and manufactured farm implements after setting up as a blacksmith in 1850. Two of his sons were sent to Edinburgh where they attained engineering degrees.

and after shearing makes a run to town. His calendar appears to circle round this epoch, and events are spoken of as having occurred so long before or after shearing. In ordinary times the station employees are few in number and vary according to the size and ideas of the owner. Two or three shepherds, invariably Scotch, and an old man or two, form the usual staff . . .

That was written in the early seventies. By the eighties, recession had set in and a pool of labour was available for the more demanding task of grain-growing. This pool of labour was seriously needed because harvesting, stacking and threshing wheat, then bagging and storing it, required even greater labour resources, especially if the farmer was fighting time – against rain or just the threat of rain.

Wheat had been grown for many years in New Zealand, at first in small fields for tribal or missionary settlement subsistence, but some was exported to Australia from the earliest times. The area under the crop in 1861 was only thirty thousand acres and falling. Gold had something to do with that at first, attracting labour away from the farms. But during the second half of the decade, stimulated by goldrush demand, the acreage began to climb. It increased between 1870 and 1880 from 86,000 acres to 157,000, and during the early 1880s to 321,000. By 1881 forty small flour mills were in operation in Canterbury, but the number diminished as large steam-powered mills took over.

The acreage was high again in the first two decades of the twentieth century and climbed again to 160,000 acres in 1920–21. Most years since then, New Zealand has had to import up to one-third of its wheat consumption and growers have had to match the high milling quality of the Australian product. Despite a serious price slump in the 1980s, a number of South Island (particularly Canterbury) growers persisted with wheat, supplemented with other grains and with grasses for seed as well as sheep. The area sown in recent years has been around 70,000 to 90,000 hectares and experienced growers with high-quality varieties had yields twice as high as before World War Two.

Grain production peaked in Otago in 1890, when 376,000 acres of oats were harvested as well as 294,000 acres of wheat. But as with wheat, yields did not hold up well until better-prepared seeds of improved strains were reared under more skilled management.

Opposite: Wheat yields in New Zealand have always been high by world standards, but farmers find it almost impossible to compete with the vast areas under wheat in Australia, North America and Russia.

Years before the wheat bonanza, visionary farmers like John Barton Arundel Acland developed one of the great early New Zealand farms, at Mt Peel, in Canterbury, and was growing substantial areas of grain. By then he had also built the Church of Holy Innocence, which still stands on the Mt Peel property.

The main crop

Grass has long been and remains New Zealand's predominant crop, the one on which its agricultural wealth depends. For nearly three decades before and immediately after Britain joined Europe the perception of experts was that food production from pastoral farming would not be enough to provide prosperity in a post-industrial world. But by the first decade of the twenty-first century pastoral farming was still the butter on New Zealanders' bread.

The earliest pioneers ploughed land to plant cereals and root crops, but as wool became the main export commodity farmers burnt and ploughed the land to sow grasses and clover from Britain. When the Liberal Government's Minister of Agriculture, Robert McNab, wrote in the 1890s that New Zealand was 'essentially an agricultural country, with peculiar advantages for the raising of stock', he was not the first to have noticed. The more intensive farming that followed the development of refrigerated shipping services in the closing years of the nineteenth century led to an increase in the ploughing of land to sow top-quality imported grasses.

At the beginning of the 1860s about a quarter of a million acres had been ploughed. A decade later, this had more than trebled; and by 1890, 7.5 million acres had been ploughed and all but one million put down in grass. The rest was in cereals and other crops, some of them supplementary fodder for pastoral animals. In no other country was such a high proportion of land cultivated for the sowing of grass; but then no other country had such suitable conditions for a long and lush grass-growing season. According to the *Official New Zealand Yearbook*, 12.5 million acres were in sown grasses in New Zealand by 1906, compared with one million acres in Victoria which had the largest acreage of pasture in Australia. In fact, the six Australian colonies together had a total of just 1.1 million acres in pasture.

As the bush was felled and burnt in the North Island and gave way to dairy cows and lambs for fattening, and as swamps were drained and scrub cut in both islands, more and more land came under the plough and was sown in profitable pasture – fourteen million acres by 1914, twenty million a few decades later.

From the 1890s the Department of Agriculture began research into the most suitable pastures not only for New Zealand's conditions generally but for the various regions and soil types. But the task of ploughing was not just a question of harnessing the team and setting off. In the open lands of the South Island, swamps had to be drained, scrub and clumped tussock cut and grubbed out, and tens of thousands of stones removed by hand. In some places dried out land had to be irrigated. The two-tined plough which appeared in the 1870s made the task of ploughing less time-consuming once land had been cleared.

For pioneer farmers, turning the forests of the North Island into pasture – from the 1880s well into the twentieth century – was often a ten-year project, frequently so onerous that many walked away from it. They would start by cutting down the undergrowth, then fell the trees

Grass is likely to be not only the staple feed of livestock in New Zealand but also the supplementary feed, in the form of hay or silage. Pictured is a primitive hay baler about 1895 with men working from a haystack to make the feed more transportable to stock in the paddocks.

in the winter, before burning as much as they could when the branches and undergrowth had dried out in the summer. Seed was sown by hand among what was left of the charred stumps and trunks. The potash left by the fires led to lush pasture, but while lambs fattened on it, their wool was torn and discoloured. It would be several more years of hard work before a plough could be used. The invention of the steam traction engine made stumping less arduous and diminished the work of hauling logs from paddocks.

The application of superphosphate became widespread in the 1920s. Superphosphate is prepared when phosphatic rock is ground into powder and treated with sulphuric acid to add sulphur and increase the solubility and consequently the ability of plants to absorb it. In 1928 a grasslands research station was established in Palmerston North and the work of A. H. Cockayne and his successor, Sir Bruce Levy, identified and isolated nitrogen-fixing clovers and a perennial ryegrass. Soil scientists researched regional deficiencies in trace elements such as selenium and molybdenum, and the development of the North Island central volcanic plateau became possible after a lack of cobalt was detected. The Northland gumland soils were also brought into production when lack of trace elements was ascertained.

The first grass mixture mostly used was perennial ryegrass and cocksfoot with some Italian ryegrass and smaller amounts of white clover and perhaps alsike and timothy. A good mixture of seed was important because the result would be the pasture the farmer would have to live with until such time as his paddock was clear enough to plough.

The research continued in a bid to help intensive modern farming live in the increasingly urban environment by developing grasses that would fix nitrogen even more successfully, thus reducing the need for fertiliser and its harmful run-off.

New Zealand was growing nearly an acre of wheat for every person in the country during the 1880s and 1890s and net exports reached 5,142,000 bushels in 1883–84. Yet the yield was low and getting lower. According to Hilgendorf:

> Wheat growing either spread into land that was not suited for it, or else the same land must have been used for wheat growing either continuously or in a very close rotation. Part of the land now occupied by Lincoln College was cropped with wheat for thirteen years in succession, after being broken in from the tussock, and indeed such a practice is hardly surprising, seeing that the farmer of those days had little other outlet for his activities. The population of the country was still small and the internal demand for such crops as potatoes trifling, but, above all, there was no frozen meat trade, and so no fat lamb industry until the mid to late 1880s. Nothing could be exported except wheat and wool, and so wheat was grown year after year. There were no manures and little rotation of crops, and so the yield, starting low, went down and down.

Yields may have been abysmal at around twenty-three bushels to the acre by 1900 compared with what the land had been capable of, but in global terms they were not low: New Zealand was still among the top half of the world's wheat producers in relation to yield.

Exports of all grains reached 13.3 million bushels in 1901, up 3.5 million from the year before, according to the figures issued in 1904 by the Registrar-General, E. J. Von Dadelszen, and then the crop declined spectacularly to less than six million bushels in 1902. In 1908 the Minister for Agriculture, Robert McNab, an historian as well as politician, wrote:

> The yields of grain are the highest in the world with the sole exception of those of Great Britain but so successful are sown grasses and forage plants and so excellent is the pasturage of the native grasses that the production of wool, meat and dairy produce has proved more profitable than grain growing and exports consist of (1) wool (2) frozen meat and (3) dairy produce; while the cultivation of wheat and oats has been reduced to the limits of domestic requirements.

The see-sawing of wheat acreage was the result of fluctuating prices, which, of course, reflected demand. World prices were below local returns, so production in excess of local demand had to accept the lower return received for exports.

So the tide turned against grain about the same time as the century did.

The New Zealand farmer developed an early attachment to machinery as acreage for crops expanded rapidly. Pictured is a harrow for eradicating weeds. Land was seriously overcropped and wheat yields were getting lower and lower by the beginning of the twentieth century. When soil was exhausted after crops in several consecutive years, harrows were used to prepare for sowing grasses.

As the fat lamb trade with Britain grew, much of the marginal cropping land was put down in English grasses and the concept of mixed farming spread. Wheat was rotated with feed crops such as rape and turnips. Many of the large stations had been subdivided and farmers with small holdings worked the land more intensively and more sympathetically. By October 1907 McNab reported landholdings of one acre or more numbered 73,367, almost double that of 1890, which was just before the Liberal Government began breaking up the big estates. Many smallholders on quality land had turned to dairying, although export dairying was still hampered by inconsistent quality because of the inadequately equipped cowsheds and factories and the varying ability of farmers and factory managers.

Wheat competed in a free market, with farmers sniffing at any suggestion of control as socialism, until 1914 when a drought virtually destroyed the Australian crop and created a crisis that forced governments to intervene and growers to acquiesce in holding prices down. From then on, until recent years, governments forced a measure of control to ensure constituents had access to the staff of life at reasonable prices. The relationship among growers, manufacturers and governments during these decades was never easy and conflict forced the marketing agencies into many shapes and forms as the market became distorted in one way or another. The end of any regulation came in the 1980s when the radical Fourth Labour Government cut farming supports. Some wheat farmers never recovered.

About fifty varieties of wheat were grown in New Zealand from when the first seed was introduced from India, through Australia, in pioneering days. Three varieties dominated the wheat fields of the nation from 1860 to the end of the century – Tuscan, Hunters, and Pearl, grown in roughly equal proportions. A new strain of Tuscan then took over more than sixty per cent of the market until coming up to World War Two when the Wheat Research Division and later the Crop Research Division of the Department of Scientific and Industrial Research began releasing local adaptations. Hilgendorf, Cross 7, Aotea, and Arawa became staple varieties, with Aotea winning more than eighty per cent of farmers' preference within two or three years of its introduction in the 1950s. During the 1970s, Aotea was followed by Kopara, Takahe, and Rongotea.

Superphosphate began to be used around 1910, and from 1920 tractors became available and better prepared the soil and provided more consistent and faster drilling and harvesting. The period between harvest and seed time, when the ground is often very hard, was short for the amount of work that had to be done with horse teams. In a capricious climate, when rain could seriously damage a crop, fast harvesting as soon as the time and grain are ripe is a huge advantage. Seed treatment for disease control and the development of varieties to suit local soil types and microclimates have all helped pull New Zealand's yield up past fifty-five bushels an acre in the best years, among the world's best after Europe where cropping was even more intensive.

Nevertheless, despite excellent and increasing yields – and techniques such as artificial drying of the harvested grain which have spread the potential land available for wheat – this country's scale of operations cannot match the enormous mechanised production from the vast, if comparatively low-yielding, areas of Australia, North America and Russia; but, more importantly, in most years there are easier ways for most New Zealand farmers to make a living.

But that isn't the whole story. Even today many farmers in Canterbury would make an economic sacrifice rather than move back into the labour-intensive growing of wheat with its need for more meticulous organisation and the risks of good and bad seasons. A researcher after World War Two reckoned that the wheat-growing area of Canterbury could be raised to 330,000 acres consistently without soil deterioration, and including other areas round the country, he claimed, the annual area could be lifted to 400,000 acres, yielding more than twenty million bushels a year. But even if the government-set price was to be lifted above reasonable expectations for fat lamb returns, it is doubtful that farmers would oblige.

What is true of the history of wheat in New Zealand is also broadly true of oats, except that they held up as an important crop past the period of wheat decline at the turn of the nineteenth century because of its value as stock food, and because it can be grown over more of the country than any other cereal.

Opposite: Early in the twentieth century, horses were being dispensed with for ploughing large-acreage paddocks for cropping as traction engines became available. Pictured is a very well-dressed man operating a multi-tine plough from behind one of the large, powerful machines in 1910.

Following pages: Even after wheat acreage declined, the demand for oats continued. A small amount was used for breakfast cereals but mostly it was a high-energy food for horses while they were the main means of haulage and transport; or it was fed out in the paddock as a supplementary greenfeed for livestock. The picture shows a field of oat stooks on the Marlborough farm of George Gunn, two of whose sons are shown with the horses. It was taken by Arthur John McCusker, of Blenheim, and it was included among New Zealand exhibits at the 1915 Panama-Pacific International Exposition in San Francisco. It won a prize and later became the image on a popular postcard.

Oats had a following from porridge-loving Otago Scots but that market was met by a small amount of production. Even today when cereal fibre is a growing diet choice, the demand can be satisfied with crops from less than 20,000 hectares, most of it in Otago and Southland. Oats retains value as a supplementary greenfeed for livestock. The highest area sown in oats was in the 1890s through to the beginning of World War One when it reached around 300,000 acres. Most of it was used as the staple, high-energy food for the horses which were the main means of transport and were breaking in the land.

Barley also had an erratic history of moving from undersupply to surplus and back again until recent years when yields and prices have led to an expansion of production, some of it for export. It was used mostly for malting for the manufacturing of beer, but crops expanded in the North Island – especially in Wellington, Hawke's Bay, the Bay of Plenty and Auckland – during the second half of the twentieth century, most of it for use as stock feed, either threshed for pigs and poultry or as greenfeed for other livestock.

Worldwide, maize ranks in importance with rice and wheat as a cereal with huge and increasing acreage and yields, particularly in the United States. When it was introduced, it was the carbohydrate plant Maori most valued, after potatoes. They soaked it in water and ate it raw, especially the young cobs, and carried the dry grains as travelling food. Maize has been mainly a North Island crop in this country, produced over many years with steadily increasing yields from about 5000 hectares – about half of it for green forage and half for grain.

An explosion of production took place from the late 1960s in Hawke's Bay, Bay of Plenty, Poverty Bay and Waikato regions, mainly in response to more productive and profitable varieties which have resulted in extraordinary yields in this country. It helped meet increasing demand for starch and vegetable oils, for poultry and pig feed, and for fattening cattle, which eat it as it stands.

Auckland, the entrepôt

While sheep dominated the economies of the southern settlements, wool was not a dominant product in the north. The wooded slopes that rose gently from the Waitemata Harbour were decreed to be a capital city by Governor Hobson in 1840; so it was immediately a government centre and within a short time it became a major trading port. After twelve months it had a population of 1500, not including the province's 59,000 Maori. The energy and initiative of the Maori down country fed the population, and whatever agriculture spread around the city was aimed mainly at supplementing this supply.

The southern settlements were angry about Auckland becoming capital and they gave Hobson a hard time. They were kitset colonies whereas Auckland was knocked up in a hurry like the raupo (bulrush) whares in which most of its people then lived. In 1845 the *Nelson Examiner* ranted in the full-throated prose of the journalism of the time:

> We have asked before and we ask again, how it comes that the government does not publish statistical returns of its own settlements? Where are we to find an account of the cattle, sheep and crops of Auckland and its dependencies? The last census in Cook Strait was taken by the government; was none taken in the north? If not, why not? If any was taken, why not published? Were they afraid of showing the nakedness of the land – of giving a beggarly account of empty boxes – of proving by comparison of figures the propriety of Hobson's choice and FitzRoy's adoption?

Some figures on the business of the ports were published by Auckland's *New Zealander* and showed that, in terms of ships and tonnage of exports, Auckland was way ahead of Nelson almost from the first day of its existence and that by 1844 it had passed Wellington in numbers of vessels, but not in export tonnage.

Auckland was even then a commercial town, an entrepôt. It was surrounded by forest, had no Wairarapa at its back door nor any huge tracts of tussock-covered plains, so sheep on any scale were impracticable. Agriculture pushed back through what is now Epsom, and crops were grown, cows milked and steers slaughtered to feed the locals, to meet the demands of new arrivals – including British troops brought in to fight against Maori – and to supply the ships that came from round the Pacific. Indeed, the budding city had closer ties with Sydney than with other centres in New Zealand. In 1848, for example, 1430 people arrived in Auckland, more than half of them from Australia (699 from New South Wales and 632 from Britain). Three hundred and seventy-two left the city (306 of them for New South Wales and twenty-five only for Britain).

Auckland didn't have the urgent need for wool to export because it had kauri gum and timber. These extractive industries certainly held back its agriculture. An Agricultural and Horticultural Society was formed in 1843 but it hardly thrived – some years, no show was held. One of the men who regularly collected a prize for sheep in the fifties and early sixties was John Grigg, the man who later moved to Canterbury and became a legendary sheep and wheat farmer. For aspiring pastoralists, south was the way to go.

Through the fifties, returns showed there were generally as many cattle as sheep in the Auckland region. Flocks were small and mostly aimed at the domestic meat market. Wool was only occasionally listed among exports.

By the eighties, the Auckland Agricultural and Horticultural Society had become, grandly, the New Zealand Agricultural Society, and then, humbly, the Auckland Agricultural and Pastoral Association, which was in such serious financial trouble that no show was held for three years until a burly Mangere farmer, Bill Massey, became president and reorganised it – a light administrative workout for a later, longtime prime minister.

In tone and character, Auckland was a metropolitan Pacific port. But in 1850 a perceptive correspondent wrote a prophetic article for the *Maori Messenger*:

> At an early period of the publication of this journal we compiled a mass of intelligence on the subject of dairy farming and endeavoured to show to what great and beneficial results it would lead if the native landed proprietors could only be brought to turn their attention to a branch of rural industry which, whilst it became the means of reclaiming and enhancing the value of their waste lands, would gradually pave the way for a great and lucrative source of immediate prosperity.
>
> The Northern Island of New Zealand is peculiarly adapted for dairy farming. Its pastures are rich and nutritious in a remarkable degree. There is ample sufficiency of moisture throughout the year to sustain and nourish vegetation, and we need but point attention to a well known fact, namely, that the Tamaki meadows have throughout the year maintained a hundred oxen on a hundred acres of ground to prove the capacity of the country, and to show that New Zealand might profitably become the dairy farm and market garden of the Australians . . . The small supplies hitherto shipped from here have realised excellent prices and have commanded the highest character. Here, then, is a mine of wealth in which our native yeomanry would do well to dig . . .

By 1852 exports of butter and cheese from the port of Auckland earned £456, half as much as kauri gum but only about one-third as much as potatoes. In the meantime, Auckland grew as fast as it needed to. Land development was expensive. It wasn't until the next century that the full agricultural and pastoral potential of the north was fully understood.

CHAPTER SEVEN

UNWELCOME IMMIGRANTS

Outside New Zealand and Australia, no creature has had a better press historically than rabbits, except maybe fairies; although it is acknowledged there are bad fairies, but no bad rabbits. They are soft and cuddly and are reluctant to bite or scratch. So why do New Zealand farmers hate them? Arnold Wall explains in his 'Song of Brer Rabbit in Central Otago':

Where the sheep fed, there feed I,
Depleted lands behind me lie,
Of dogs and guns I take no heed
I only breed and breed and breed.

At traps and guns and dogs I smile,
I laugh at cats' and weasels' guile,
I frolic, nibble, frisk and feed,
But all the time I breed and breed.

They send my skin to clever folks,
Who turn me into sealskin cloaks,
A sorely hunted life I lead,
But still I breed and breed and breed.

They chase me round with mongrel packs
And bear me home in bulging sacks:
I smile, and say my simple creed
Which is: 'I breed and breed and breed'.

When Nature gave the brutes their law
To some she gave the tooth, the claw,
The poison-fang, the lightning speed;
Me she bade only, 'breed, breed'!

When foes attack I cannot bite,
I have no spirit for to fight,
By other methods I succeed
I merely breed and breed and breed.

The Nation which would keep her place
And beat her rivals in the race
This little ditty first should read
And then go home and – follow my example.

No one is quite sure who introduced rabbits into New Zealand but it was clearly more than one person on more than one occasion. A pair of rabbits was among the prizes for a ploughing competition at Waimea in 1843. The records of the 1849 Auckland Agricultural and Horticultural Society mention that, among the livestock, 'Dr Courtenay exhibited a remarkably fine pair of rabbits.' It was part of the sentimentality of the British immigrants trying to transplant familiar plants and animals – the same sentimentality that brought in other exotic mammals which turned out to be costly pests and that gave gorse and other noxious weeds a foothold. The importation of rabbits was especially ill-considered because their depredations in other lands, notably Australia, were already known by then; but it was understandable because the ordinary Englishman and Scot knew the rabbit only as a useful, controllable animal in his homeland. Many of them brought out other plants and animals to this country's enormous benefit.

Gorse became a scourge as it spread like a yellow flame through the country, claiming land from

A well-equipped rabbit hunter on the lookout, with his pack of dogs and rifle, in scrappy country. They couldn't shoot them fast enough.

grass, but at least until the turn of the century it had its supporters – and official supporters at that. The Department of Agriculture's officer in charge of the Auckland Province in 1897 said in his annual report:

> Gorse-growing for feeding sheep is being considerably extended in the Bay of Islands. The energetic owner who initiated the growing of gorse for fodder is gradually overcoming prejudice, and the next step will be others following his methods. This is a most interesting development in utilising country that until now has been comparatively wasted, and if an experimental station is established here one of its most interesting features should be the number of sheep per acre grazing in its gorse paddock.

The first few rabbits brought in either didn't survive or didn't do well enough to spread beyond local colonies. Rabbits were undoubtedly around this country for more than twenty years when, writer and historian R. M. Burdon reported, 'one of the passengers in an immigrant ship that arrived at the Bluff in 1864 . . . brought out with him some rabbits as a speculation. He sold them at high prices and they were turned out in the vicinity of Invercargill.' They dug in first on the sandhills near the town but it appears they took some time to adapt to the nearby heavy country. Once they hit the hills, though, and the drier country in the middle of the province with its light, sweet native pastures, they found their Eden.

Something of a lemming-like inexorability drove the voracious creatures northwards, accompanied by a locust-like way of eating out everything as they went – everything a sheep would want to eat, anyway. They grazed pasture right down to the ground, closer than any other grazing animal, and the leaf was sawn off again at the shoot as it reappeared. So sheep went hungry. In the hot, dry areas

A Waikato hillside covered with a tight mat of gorse. A senior Department of Agriculture officer wanted research done to discover how many sheep per acre could graze a 'gorse paddock'.

Opposite: Rabbits did not just compete with sheep for pasture, they destroyed it as in this paddock that resembles a moonscape.

with light soils, the surface would shift in the autumn with no carpet to hold it in place. Goldminers helped transport them round the province by releasing pairs near the diggings to breed fresh meat. That's the way they got over the Clutha River.

In 1876 a government committee of inquiry was set up as a result and the same year the Rabbit Nuisance Act was passed giving responsibility for control to local organisations. It didn't work because some farmers weren't as badly affected as others and the rabbit had a refuge if they didn't try to hunt him out. It was a bad year south of the Waitaki, as Burdon pointed out:

> In 1876 on one run where three years before hardly a rabbit could be seen, sixteen men and one hundred and twenty dogs were employed in keeping them down. They killed 36,000 in one year; but the numbers did not seem to be diminishing. On another run 50,000 had been killed; but the average weekly expenses of £27 were fast ruining the owner. On the plains men hunted them down with dogs; on hilly or broken ground, they shot them. Some went in for trapping and others tried poisoning, but this was soon given up, because the hawks fed upon the carcases and died, and they were looked upon as the rabbits' greatest natural enemy. In parts of Southland, where the increase of lambs had been sixty-five and seventy per cent, it fell to twenty. The yield of wool for the province fell by eight hundred bales . . .

The furry plague picked out the warmest and sweetest country and destroyed it, and spared only that which was damp and unhealthy. Runholders in some parts of Central Otago walked off their land and effectively turned it over to the rabbits. Much of Central Otago became a disaster area and some of it has never recovered. One of my earliest memories is driving a few kilometres back from Dunedin one night in the 1930s with my father and a friend on a hunting expedition. They shone a strong light up from the road and the hillside seemed to move as the uncountable rabbits raced for shelter. Rabbit was tasty and tender meat, sold skinned and dressed at low prices in fish shops. A magazine description of Central Otago, published in 1882, read:

> Hills and gullies that used to present a scene of perfect sylvan beauty, with variegated

pasture intermixed with sparkling streams and alpine snowcapped ranges, and literally covered with sheep, now look like a deserted waste, as though some gigantic deluge had swept vegetation off the earth.

Another formal inquiry in the 1880s resulted in more legislation – this time more determined and centrally directed. But the rabbit spread into Canterbury in numbers, into Marlborough, and then infested parts of the North Island: Wairarapa, Hawke's Bay, Poverty Bay and the centre of the North Island. It thrived anywhere the land was light and dry and the pasture not too rank.

The problem was nowhere else as bad as in Southland and Central Otago. A measure of control was gained in most areas by the end of the century through the use of poison, mostly laid with pollard, a fine bran sifted from flour, but an impediment which persisted where the rabbits were thickest was their commercial value. The Department of Agriculture's assistant chief inspector of stock reported in the closing years of the 1800s that the trapping of rabbits for freezing and export would, in his opinion, become an industry of 'large dimensions'. He wrote:

> Packing houses have been put up at many railway points, and at these houses the rabbits are packed in cases, twenty-five and thirty in a case, and the cases are then trucked to the freezing works. This freezing industry is a difficult matter to deal with. The rabbits are not caught for the sake of decreasing their numbers, but simply because they realise a profit. What must be guarded against is that rabbits are allowed to be farmed for this special purpose. In some parts of the country I found that very little work was being done during December and January in the way of keeping down rabbits, pending the trapping season opening in February. This was deliberate rabbit-farming and cannot be countenanced.

It is not hard to understand that a farmer – whose ability to earn money from sheep had been severely curbed by rabbits – would want to get some money back from the enemy. In 1893 seventeen million rabbit skins were exported and earned £139,000, and that same year 100 million pounds' weight of frozen rabbit meat was exported. The number of skins declined from that level down to only five million in 1906, but the volume of frozen rabbit meat grew to 226 million pounds that year and stayed at around 190,000 pounds until World War One. The skin export trade burgeoned again to nearly nine million in 1910, worth £130,000, and stayed around that level until the war. These values were minuscule compared with the export returns for wool, but because of the trade in rabbit skins and meat the country was unable to get on top of the problem until after World War Two when the government took legislative steps to decommercialise the rabbit. Not that the problem went away, but it did diminish somewhat.

The insidious nature of rabbit infiltration, invasion and domination was captured dramatically by R. M. Burdon in his 'Canterbury Under Siege' chapter of *High Country*:

> Between Canterbury and Otago nature has placed a very effective boundary in one of the South Island's greatest rivers. From the three glacier-formed lakes of Ohau, Pukaki and Tekapo emerge three rivers which have their confluence some miles to the east, where they form the Waitaki river. Its swift, icy waters might have been expected to provide a barrier impregnable to creatures so puny as rabbits; but they had crossed large rivers before, and by 1887 had begun their attack on the right wing of Canterbury's main line of defence. They passed over the mountainous country between Lakes Hawea and Ohau, where they divided into two armies. One debouched on to the arid Waitaki plains and made its way eastward along both banks of the Ahuriri; the other, moving round the northern shores of Ohau, crossed the Hopkins and Dobson rivers and threatened to cross the Tasman. Further to the south by some means or other they crossed the Waitaki and appeared on the Waitangi and Tiakatarawa stations; and it became apparent that unless they were

stopped the whole of the hill country between the Waitaki and Rangitata rivers would soon be overrun.

Simultaneously the attack was developing from another direction. Three hundred kilometres to the north, rabbits were pouring into the Amuri district from the province of Marlborough. Canterbury was beleaguered. The position was so grave that the government – taking a leaf from the book of the Chinese, who built a wall hundreds of kilometres long to keep out the northern barbarian – put up a wire-netting fence, 128 kilometres long, to keep out the southern pest. Something of the sort had already been done in the north, where the Hurunui rabbit fence, beginning at the sea, ran inland for eighty kilometres. Caretakers were employed to patrol the fence to destroy any rabbits found within a kilometre of it on the Canterbury side; but in spite of these precautions a few got through and reinforced those that had made it in before the fence was constructed.

Even more eloquently, the impact of the unstoppable migration is described by that indefatigable observer of nature in Hawke's Bay Herbert Guthrie-Smith, in his 1920s classic *Tutira*:

> Rabbits turned out in the Wairarapa after a few seasons became a curse, increasing and multiplying until many of the local squatters were eaten out, their stations left desolate, and they themselves ruined. Between that period and the date of their invasion of Tutira the hand of man had laid heavy upon the rabbit. Everywhere also its advance had been retarded by the attacks of harriers, moreporks, and wekas, working not for eight hours a day like man, but for twenty-four. Rabbits nevertheless have forged ahead thinly on a vast front, reaching Tutira almost simultaneously by the hill-top, by the coastal route, and by the line of the human highway. There was no strip of open land from coast to mountain ridge along which

Hawke's Bay naturalist Herbert Guthrie-Smith watched and noted their progress as rabbits plagued his famous property, Tutira, the subject of one of New Zealand's classic works of literature.

during the 'nineties rabbits were not filtering towards Hawke's Bay – towards Tutira. It was an invasion which, however disgusting to the sheep-farmer, was full of interest to the field naturalist.

The physiography of Tutira has been described – a series of slopes precipitous to the west, tilted gently to the east, and fissured by ravines. It was a pattern well adapted to illustrate the rabbits' desire to follow a definite direction. The invasion moved from south to north, the ravines on the other hand ran east and west, at right angles, that is, to the line of trek. On the high ranges of the west these natural obstacles to progress were particularly formidable. On many the gorges began within a few chains, sometimes within a few yards, of the ridge-cap. Only rabbits, therefore, moving along the very summit could proceed. All others found their progress barred, discovered themselves on the brinks of precipices, many of which were fifty or sixty feet in height.

It was impossible for them to remain for any length of time on barren cliff-edges. By natural enemies, necessity of food-supply, fear of the open, and lack of burrow accommodation, they were eventually forced either once more to the summit or downwards to more accessible country. Thus barred from their selected line of progress many were forced downhill, and reached after a time the great pumiceous trough of the run. Through this, at that time, bleak forbidding country ran east and west the main stock-route of the station, the trail by which sheep were driven to the distant paddocks and mustered homewards for shearing. It varied in width from five to ten yards, and for miles passed through fern-lands bare of grass. Now it might have been anticipated that rabbits thus finding themselves on an open road, a smooth space hedged in by deep bracken, would have been content to accept it without question. Far from this happening, however, the east and west stock-route was used no longer than was necessary to find

Millions of rabbit skins were exported from New Zealand each year until the end of World War Two. Schoolboys were happy to earn money stretching skins for drying in the open air, as shown here in Petone in 1932. The skins were cleaned, stretched, dried and then baled for export. In the late 1950s rabbits were decommercialised because it was felt that the industry encouraged keeping a supply, thus working against the campaign to keep their numbers as low as possible. Although overseas earnings from rabbit meat and skins were significant, they did not equate with sheep grazing the same area, and they ruined pasture by nipping grass down to its roots.

Horse and dog

The hackyard at John Grigg's Longbeach farm in Canterbury.

Two imported animals that proved to be of enormous basic value to farming are the horse and the dog – although there was a period last century when wild dogs were so feared as predators that farmers gained the right to shoot on sight any unregistered dogs on their property.

The first dogs came with Maori. The kuri was described by Frenchman Julien Crozet in 1772 as 'a sort of domesticated fox, quite black or white, very low on the legs, straight ears, thick tail, long body, full jaws, but more pointed than those of a fox, and uttering the same cry; they do not bark like our dogs'. Dogs arrived again with the sealers and whalers and the pure kuri became extinct. The Scottish shepherds who came to New Zealand from the 1840s often brought their dogs with them, and these and other imports since have combined to make the New Zealand farm dog. Just as the shepherd's dog became an asset at mustering time, the wild dogs that lived in the bush became a menace to the flocks, and to this day farmers have an enduring affection for their work dogs but an abiding hatred of the wild predator.

In pioneering days, dogs were used more to contain sheep in the absence of fences than to drove them. The shepherd more often led his flock with a bellwether. Dog-training developed as a work-related pastime – a sport in its highest form. Animals like Border collies are intelligent and quick to learn, and farmers soon came to understand that dogs moving to an elementary set of commands were a huge asset at droving or mustering.

The horse arrived with Marsden, and there have been large infusions of blood since then to build up the stock of draught horses, station hacks and, lately, sleek and highly strung racing animals. The horse reached its peak as a work animal in the first quarter of last century and was replaced by first the traction engine, then tractors and motor cars. During the oil shocks of the 1970s some farmers claimed the horse could make an economic comeback if energy costs rose high enough. The factors that count against this renaissance are the high intake of food by the working horse – more than twice that of a cow – and the time and cost of preparing it for work and rubbing it down afterwards, and of caring for the harness.

Experience dictates that no one should bring any exotic creatures into New Zealand in the twenty-first century because the dangers for the environment could be at least as devastating as the introduction of rabbits and possums – except that in recent years government agencies, after exhaustive research, have brought in what they consider host-specific insect predators from time to time to control such pests as grass grub. In the age of rapid transport moving hundreds of thousands of people and millions of tonnes of cargo throughout the world each year, animals the size of possums may be capable of being controlled with some ease compared with insects and, worse, diseases. Biosecurity services now attempt to prevent potentially harmful organisms from coming in here through border-control services based on carefully calculated risk-assessment procedures around the country, backed by offshore inspections, surveillance, and quarantine techniques and facilities.

northward-leading exits. No pig-trail, however narrow, no sheep-track, however overgrown, no fence-line, however little trodden, that branched off northwards, was neglected. Rabbits caught by the contractor and his trappers were found in the extreme northern points of the blind spurs.

Guthrie-Smith was a rigorous observer and recorder of change in his physical environment, the outstanding amateur naturalist of his time. As he went about his farming work, he also noted the arrival and settlement in Hawke's Bay of hundreds of exotic plants and many birds and beasts.

Poisoning and trapping rabbits, often by contractors employed by farmers, kept a measure of control but, of course, it wasn't in the interests of those making money from skins and carcasses to let breeding stock numbers fall too low. In 1873, 33,000 rabbit skins were exported, about a million in 1877 and more than nine million in 1882. By 1894 exported skins totalled 17,666,460, and in 1901 six million frozen carcasses were shipped out in a rapidly expanding new trade. So it wasn't until trading in skins and carcasses could be stopped that the rabbit threat to sheep farming began to dwindle.

It took a long time, though, to kill the trade in skins. According to P. G. Stevens in his book *Sheep*, 134,000,000 rabbit skins were exported from New Zealand during the ten years up to 1949, a figure that will astonish most people who thought the rabbit menace was diminishing sharply during the 1940s. There were those who considered rabbit exports so valuable that they should be left to earn money for the country, but the cost far outweighed the gains. The rabbits severely damaged the light-soil pastures on South Island hills and the subsequent erosion damage was often irreversible. Experiments showed that ten rabbits displaced the grazing of one ewe but were far less valuable to the farmer or the country even on the basis of immediate monetary return.

By the late 1950s the decommercialisation of the rabbit had reduced it from a scourge to a pest. The campaign was backed by a heavy eradication drive and a claim, which may have been spurious, that hydatids was widespread among the wild population – an allegation that successfully discouraged would-be consumers from eating them. The more intensive stocking of hill country in response to aerial topdressing and the aerial dropping of poisoned baits also had an effect. Nevertheless, the feral rabbit will never be eradicated and sudden build-ups in population in some areas come and go like a plague. An attempt in 1951 to introduce myxomatosis into New Zealand to decimate the rabbit population failed. The disease had wreaked havoc among Australian rabbits. One reason it didn't take here is believed to be the lack of biting insects.

In July 1997 the government turned down an application to import the calicivirus disease, having taken scientific advice on two aspects of such an introduction: its effect on the ecosystem and other species, and its effectiveness – or lack thereof – against feral rabbits. What followed was an extraordinary disregard for the law when the calicivirus was brought in illegally by a group of farmers. The government not only never discovered or charged any person but retrospectively condoned the action. However, a few years later rabbits seemed to have made something of a comeback and the effectiveness of calicivirus was seen to be short term.

Rabbits successfully adapted to New Zealand conditions, but over many years attempts – happily unsuccessful – were made to introduce the Canadian moose and the Indian grey mongoose, as well as kangaroos, alpacas, gnus, chipmunks, raccoons, bandicoots, squirrels and even zebras.

The other major animal pests that survived and thrived are deer, possums and goats. As with rabbits, they were introduced with the benign intent of providing game in a country without any mammals of its own. All three creatures encountered no natural enemies and caused erosion in areas where they flourished by eating out grasses, young trees and shrubs, and defoliating larger plants in areas where their population build-up put pressure on food resources. The government was forced to send in shooters against deer and goats.

The major animal pest in the early twenty-first century, the possum, is a persistent and even more unstoppable invader of regions than the rabbit. A nocturnal animal originally imported from Australia, it is heavily represented in road kill,

'Captain Cookers'

During the centuries in which Polynesian explorers ranged around the Pacific, the domestic animals they took with them were the pig, the dog (kuri), the fowl and the rat or kiore (probably by accident). But for some reason still not fully understood, what are believed to be successive canoe arrivals in New Zealand brought no pigs with them. One suggestion is the journey was so arduous the pigs didn't survive or were eaten as the explorers ran short of food; but they did make it in expedition canoes across other long stretches of ocean, including to Hawaii from the Marquesas or Tahiti or both.

So pig meat remained off the menu for Maori until 1769 when French explorer de Surville arrived in the *Saint Jean Baptiste* two months after James Cook first rediscovered New Zealand in the *Endeavour*. De Surville traded a boar and a sow with Maori for fish and vegetables in Doubtless Bay. During Cook's second visit, a boar and two breeding sows were given to Maori in Queen Charlotte Sound. And so pork again became part of the relished diet of Maori, hundreds of years after they had left them behind in their Hawaiki. More pigs were brought into New Zealand in the early years of the nineteenth century; but because history is written by the winners, the progeny of all the early porcine arrivals were tagged 'Captain Cookers', not de Survilles. The name Captain Cookers has survived in association with the razorback wild pigs – lean, black aggressive animals often with tasty flesh and trophy tusks that lure many recreational hunters.

Organised pork and bacon production grew as a corollary to dairying and grain-growing. Perhaps because no steady profitable market for pig meats emerged, pig farming mostly remained a sideline – a way to make use of skim milk and whey rather than ditch it into rivers or streams, and for arable farmers to get rid of surplus or spoiled grain and potatoes. The quality of the stock was neglected despite imports from Britain and Australia of Berkshires, Tamworths, Large Blacks and Yorkshires. However, because of the ready availability of food most dairy farmers had a few sties; so numbers grew, and by the end of the nineteenth century the pig population was estimated at 250,000.

No one seemed much interested in breeding up higher-quality animals until the 1920s when erratic prices for dairy products saw the start of an export trade. This prompted the Department of Agriculture to set up a pig-husbandry advisory service. Pork and bacon have long been popular foods in Western countries, but working against an export trade for New Zealand was that, along with chicken, pigs could be produced in intensive feedlots near densely populated urban areas, and that they thrive on waste food, other cheap fodder and grains.

A herd-testing association was set up in the late 1930s and some progress was made, but pig-meat production faded again as dairy farmers increasingly supplied wholemilk to the factories. Nevertheless, a National Pig Breeding Centre was established in the 1950s and set up a successful performance-testing scheme with Berkshire, Large White and Landrace boars and sows to ensure good-quality production for the domestic market. Pig-meat production became a venture in its own right based mainly on maize and barley – no longer a by-product of dairy farming. The breeding centre became associated with Massey University in Palmerston North. Pork, ham and bacon production and price were hugely influenced in how they were delivered to the consumer by the advent of supermarkets.

Pork and bacon have been imported on occasions over the years when local production has proved inadequate, and in the early years of the twenty-first century the domestic price was driven down by low-cost imports (mainly from Australia and Canada). This brought about a decline in local production. As small operators withdrew from the industry, the herds of the remaining producers did get larger in search of economies of scale. The number of farms declined from about 450 to fewer than 360, mainly sited in Canterbury and the Waikato. Modern pig farming was dramatically different from the old open-paddock sty system. The intensive, factory-style farming has aroused the ire of animal health observers and forced the industry to take notice.

Whey or skim milk being fed to pigs. Commercial pork and bacon production developed as by-products of dairying and grain-growing.

Raising pigs under indoor battery systems has raised the ire of some animal rights people, but these pigs enjoy relative freedom on Ian Jackson's free-range piggery, Havoc, near Timaru.

blinded by headlights as it continues its inexorable spread throughout the country. It populates forests everywhere and in many provincial cities invades suburban copses, and may live and breed in the space between ceiling and roof. It defoliates trees, can even carry tuberculosis, and is hardy in the face of all measures to exterminate it. At various times, possum skins have had small export booms but the quality varies and commercialisation will never impact on the feral population. Large-scale poisoning campaigns, often involving airdrops, have drawn protest by those who believe other animals are affected because the poisons used are indiscriminate.

To get something useful from this notorious pest, hides are used in high-fashion garments, promoted as 'cruelty-free' fur, and the plucked fur is used in a fibre mix with merino wool to make accessories such as socks and scarves.

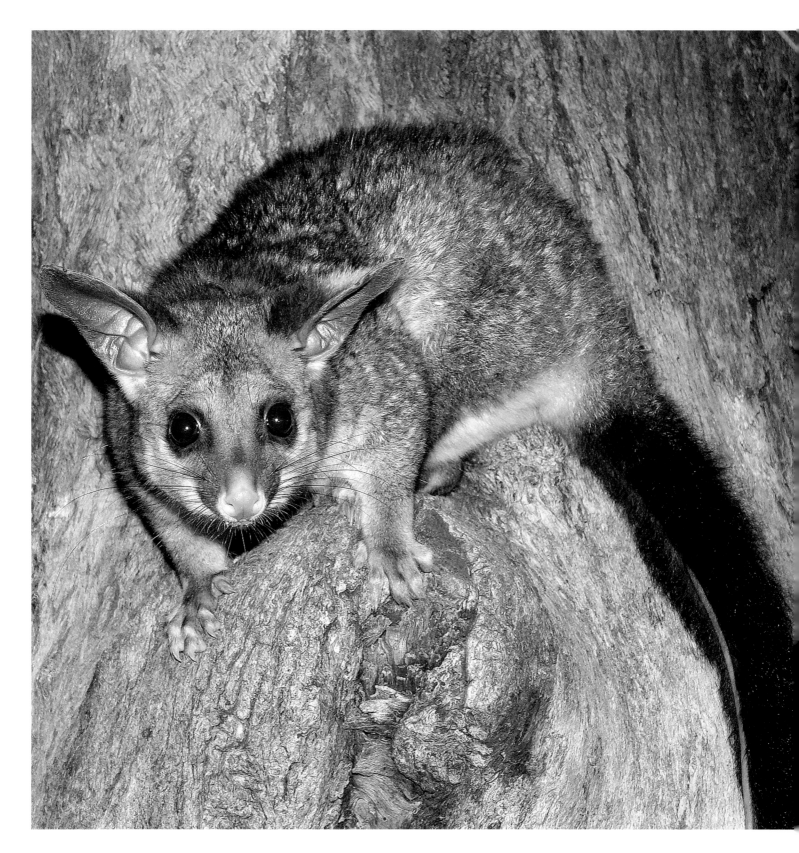

CHAPTER EIGHT

REFRIGERATED SHIPPING

THE INVENTION and development of refrigerated shipping was the greatest ever single boost to the New Zealand economy. The lucrative export conduit for meat and dairy products from 1882 lifted the young colony from the despair of economic depression. A new prosperity gradually arose from it as well as the confidence and optimism that expressed themselves in the fifteen prosperous and boisterous years following the election of the 1891 Liberal Government.

The country had undergone explosive expansion during the 1870s, thanks mainly to the massive borrowing for public works and immigration through a plan devised by Julius Vogel, Colonial Treasurer in the administration of William Fox. The government borrowed £20 million in just over a decade. Population almost doubled from 237,249 in 1869 to 463,729 in 1879 – 32,000 migrants in the peak year, 1874. A works programme saw the construction of 1800 kilometres of railway, hundreds of kilometres of road (including bridges), port facilities and public buildings, 6000 kilometres of telegraph line and an undersea cable to Australia.

This extended infrastructure opened up millions of hectares of land to production – land previously too remote to settle or from which to move produce to distribution centres. Along with the increase in stock numbers and enlarged areas of cropping came fencing, as wire and iron standards became available, giving farmers more control over stock. Export earnings from wool were strong and receipts from gold still fairly buoyant.

But with the 1880s came the country's longest economic depression in its history as wool prices dipped. In 1860 wool fetched sixteen pence a pound, then eleven pence in 1871, and, after a small boom in 1873, it kept trending down to five pence a pound by 1900. Total export returns for wool stayed about the same during the 1880s but from massive increases in volume. Gold gleamed less brightly in the rocks and riverbanks of Otago and Westland, and the glowering national debt left by the Vogel plan obscured the future of the brave new colony. For the first time, immigrants became emigrants, those who could afford it leaving for the Australian colonies or the United States, and others asking for government help to do so. Not granted. Those settled on smallholdings may have had adequate food but no money as the economy seized up and work on railways and roads became scarce.

Meanwhile the flow of settlers onto the land continued unabated. Stock censuses taken in 1871 and 1881 showed sheep numbers increased from just under ten million to thirteen million, and to eighteen million during those hard 1880s. Cattle numbers jumped from just under half a million in 1871 to 700,000 ten years later and to 831,000 by the end of the 1880s. The number of horses almost exactly doubled to 162,000 between 1871 and 1881 and increased again to 211,000 by 1891. But the value of both sheep and cattle depreciated sharply as the local market for meat was met by farms adjacent to towns and cities.

Frustratingly, the population half a world away in

Refrigerated shipping ensured that New Zealand farmers would spend the next hundred years concentrating on the production of meat and milk as well as wool. Tough, durable drovers with their horses and dogs whose job was to move livestock around the country became a common sight on country roads and in country towns before trucks took over.

Europe was surging. In England it had more than doubled from around fifteen million early in the nineteenth century to thirty-five million at the beginning of the 1880s, without anything like as much growth of food production. The impossibility of transporting fresh perishable food across the world by sea had challenged scientists and engineers in a number of countries for half a century. Americans had successfully shipped meat across the North Atlantic packed in ice but this was not an option for southern hemisphere countries whose ships had to pass through the tropics. The concept of mechanical refrigeration arose from some experiments by the great English scientist Michael Faraday, but developments were gradual. In the meantime, entrepreneurs in the countries with large production surpluses were exporting meat, mainly in three forms: preserved by boning and then drying, sealed in tins or jars, or set in barrels of tallow.

More than 200 patents were issued for meat-preserving techniques during the nineteenth century, including methods that involved the immersion in or the injection of chemicals. In Australia the Dangar brothers began tinning meat in the 1840s, and twenty years later Australian meat in soldered tins was selling well in Britain. Many British consumers baulked at the appearance and palatability of preserved meat, aware from painful experience of the poisoning potential of the product when it went bad. By 1870 New Zealand meat was being canned with only limited success, so relatively little progress was made with exporting, although technological advances in canning were being made and it was the most common preservation method used here. Immersing boned meat in tallow was less favoured than in Australia, despite the fact that preserving birds inside fat was an old, tried and successful Maori method. Prolific Otago chronicler

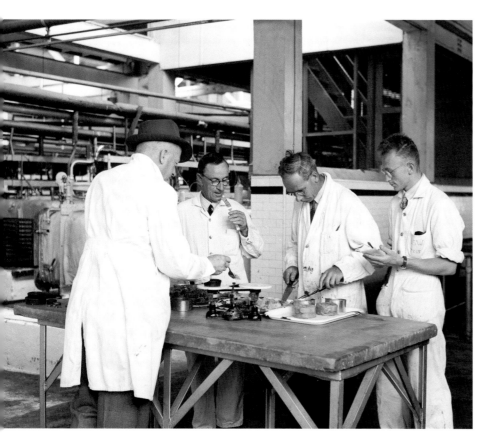

Most countries trying to export meat before refrigerated shipping became available in 1882 experimented with canning, and New Zealand did so with only limited success. However, as technological advances were made, canned meat products became acceptable to consumers and held a niche in both the domestic and export trades. Pictured is a panel testing canned products at the Westfield Freezing Works, Auckland, immediately after World War Two.

Alexander Bathgate wrote in 1874, 'The low price of stock caused attention to be turned to meat preserving, and there are now several large establishments in different parts of the province turning out by the ton this wholesome and delicious article of food. The value of such meats exported from the colony in 1872, being mostly from Otago and Canterbury, was over £160,000.'

Bathgate said that at the end of the 1860s cattle had been unsaleable in Otago but the establishment of the preserving factories had re-established some confidence and 'fat cattle were quoted at £6.10s in the Dunedin market in the month of August 1873'.

Most of the factories had been established at points around the South Island by the New Zealand Meat Preserving Company from about 1870 but, according to historian R. M. Burdon in *High Country*, many were closed by 1876 at the end of a flurry of business associated with the Franco-Prussian War. Tinned meat did have brief revivals. Twenty years after refrigerated exports began, the Department of Industries and Commerce reported that the New Zealand Agent-General in London suggested enquiries should be made within the colony on whether preserved tinned meat could be manufactured to supply British army units in South Africa. The department said it managed to negotiate an order from the Imperial authorities for 100,000 pounds of tinned meat from Mr W. A. Thompson of Woodlands, Southland. The department reported, 'A very much larger quantity could have been placed but only three offers were received in reply to inquiries, no offers being received from the larger meat-preserving works, who generally replied that their surplus output was already under offer to Australian and London merchants in anticipation of the War Office contracts.'

So, in 1902, preserved tinned meat was still good

business for certain niche markets. (Huge volumes of oats and hay for horses were also sent to South Africa during the war there.)

However, New Zealand was curiously slow in its response to the possibilities of refrigeration. Australian engineers had been wrestling with the problem for years. Thomas Mort and James Harrison had built a freezing works in Sydney in 1861 – the first in the world. Mort patented his ammonia-absorption refrigeration process in 1867 and in 1873 successfully demonstrated it in Sydney, freezing beef, mutton, fish and poultry. But when they sent their first trial shipment to London in 1876 it failed. The meat was inedible on arrival. The following year the first successful cargo of frozen meat was shipped from Buenos Aires to Le Havre, followed by a successful shipment of forty tons of edible beef and mutton from Sydney to London in 1878.

It may have been slow coming but this one technological stroke would rescue New Zealand from the prospect of perpetual peasantry, or perhaps

Solo killing at abattoirs was the way it was done before freezing works with each butcher killing, pelting, and gutting a sheep on his own. When the demand rose for carcasses to pack quickly in ships' holds for freezing, the tempo was picked up by a chain system in which each butcher filled one role as the carcass moved along.

Right: The manager of the New Zealand and Australian Land Company, William Saltau Davidson, masterminded and provided the capital for the first successful export of frozen meat from New Zealand to Britain.

Far right: A poster showing the *Dunedin*, which took the first frozen meat exports from Port Chalmers to London, and some account sales.

Opposite: New Zealand cattle and sheep have had to learn to graze on steep terrain like this majestic sweep from flats to rocky hilltop.

a kind of latter-day feudalism. Although Julius Vogel's borrowing had left the country deeply in debt, the infrastructure it bought helped give the economy a major surge as soon as refrigerated ships became a reality, because access from farm to freezing works to ports was much easier than it would have been a decade before. The man who made the first constructive move here was William Saltau Davidson, manager of the New Zealand and Australian Land Company. He persuaded his firm to invest £1000 in a trial shipment aboard the *Dunedin*, a 1200-ton, fast sailing ship, operated by the Albion Shipping Company. A killing shed was set up by the company near Oamaru, and when the *Dunedin* arrived at Port Chalmers in the spring of 1881 butchers starting killing early each morning. The meat was railed to the ship, frozen between her decks and stacked in the hold.

During the ninety-eight days at sea, the ship's master, Captain Whitson, had to contend with a fire in the sails ignited by the machinery and, when the cold air failed to circulate adequately as the vessel passed through the tropics, he crawled into the hold to cut new air ducts. He was overcome by the cold and would have died there had not the first officer crawled in behind him, fastened a rope to his leg and pulled him free.

When the *Dunedin* arrived in London, she was met and her cargo inspected by Davidson, who had ceremoniously stacked the first carcass aboard the ship in Port Chalmers and then immediately left for London. The £1000 investment proved worth every penny: the 4908 prime sheep and lamb carcasses sold within a fortnight for £8000, more than twice the price they would have fetched in New Zealand. The cargo included 2226 sheep tongues, twenty-two pig carcasses and a small consignment of butter. In June the same year the *Mataura* sailed from Port Chalmers with the second refrigerated cargo, again successfully. A year later seven kegs of butter went with the first frozen shipment of meat from Wellington aboard the *Lady Jocelyn*.

The trade boomed from the start. A number of steamships were soon fitted with refrigeration holds and some new vessels were designed and built with refrigeration equipment. It was a feverish time in the shipping business, anyway. During the last third of the nineteenth century world trade doubled in volume and the tonnage of the international merchant fleet about doubled also. It was the beginning of the era of the iron steamship.

Romney sheep have dominated New Zealand flocks for many decades and still made up half the total sheep population at the beginning of the twenty-first century despite the importation of big, fecund European breeds.

What is often forgotten is that refrigerated shipping imposed demands on the Railways Department, which responded remarkably well – at least at first. Thirty-three refrigerated trucks were rolling by March 1883, with fifty-three more within the following three years. Squabbles arose during the following decade, however, about the number of trucks on the various routes between farming regions and the major ports.

According to P. G. Stevens, the biographer of influential Canterbury farmer John Grigg, 'the first man to eat Canterbury lamb in England' was Grigg's son, J. C. N. Grigg, at Jesus College, Cambridge, along with members of the college rowing eight. Grigg had arranged the shipment, aboard the *Dunedin*, of three of his Shropshire-cross lambs and two wethers among all the New Zealand and Australian Land Company carcasses. Grigg the younger, according to Stevens, collected the lamb, then thawed, cooked and served two cold hindquarters to the oarsmen. They loved it, Grigg is said to have reported, and most had three helpings. As soon as they heard of the success of the *Dunedin* venture, Grigg the elder and a number of name Canterbury runholders with large properties got behind the new trade and formed the Canterbury Frozen Meat Company. Freezing works sprang up around Canterbury and Otago, and in those parts of the North Island with large sheep populations. Within ten years of that first voyage, seventeen freezing works were operating around the country with a capacity of 3.5 million carcasses a year. By 1911 works numbered thirty-one. Beef and pork were also being slaughtered for export, although trade in pig meat never developed.

Initially, frozen meat was mutton rather than lamb or beef. The merino, with its dark and stringy carcass, became a problem. Now that the *meat* had become such an important export, the merino's days were numbered as the predominant New

'The best farm in the world'

John Grigg was New Zealand's most famous nineteenth-century farmer. His farm was so much his alter ego he was referred to almost exclusively in one breath as 'John Grigg of Longbeach'.

A family tradition was that the Griggs descended from two Scottish brothers who migrated to Cornwall when the McGregor clan was dispossessed of its land and civil rights in the seventeenth century. John Grigg was born on the family's Brodbrane farm, near Duloe, in Cornwall in 1825. He could have stayed on the family property and lived something close to a manorial life but chose to sell up and sail for the colonies. His decision was prompted by a desire for adventure, garnished by the departure for New Zealand of Martha Vercoe with whom he had fallen in love.

John Grigg of Longbeach.

Grigg left England in 1854, stayed for a few months in Melbourne, sailed for Auckland, leased a farm at Otahuhu, and in June 1855 married Martha Vercoe. In Auckland he successfully grew potatoes for export (after one early failure), and profited from hay sales for army horses as British soldiers arrived for the wars in Taranaki and the Waikato. He imported stud sheep, and by the time he decided to move to Canterbury in 1865 had built up a flock of pure-bred Leicesters that drew attention from throughout the colony when it was dispersed at auction. He was a lay reader and one of the signatories of the original constitution of the Church of England in New Zealand.

Grigg decided to acquire a property already known as Longbeach where the Hinds River ran from hills to the west across shingle beds and then disappeared into a large swamp before it reached the coast; on the northern side of Longbeach was the Ashburton River. Surveyors called it an 'impenetrable swamp'. Graziers ran a few cattle over the drier parts but otherwise kept clear of it.

John Grigg freeholded 2000 acres in 1864 and, by 'gridironing' – purchasing strategic blocks within a larger area to render small blocks uneconomic for others to retain or buy – bought 32,000 acres over the following seven years. He moved south in 1866, bought a house for the family in Christchurch and then lived in a sod hut as he audaciously dug a bed for the Hinds to run into the sea, thus beginning to drain the swamp. It was a remarkable feat that took a long time, but he once said, 'On Sundays I used to go and lie down on the beach, listening to the sound of waves and dreaming of what Longbeach should one day be.'

By 1880 the farm was stocked with 10,000 merino and half-breed ewes using Leicesters and Lincolns and even the special mutton breed, Shropshires, as he set out to breed sheep to serve the British meat market before refrigerated shipping techniques had been proven. Grigg was an enthusiast for pure-bred stock. He had Clydesdale and thoroughbred studs among four hundred horses on the property, and sold about eighty horses a year. The farm also carried more than 2000 pigs.

Meanwhile, Longbeach had a cropping programme which had begun in the 1860s. By 1879, 3000 acres were under wheat, oats and barley, 1400 under roots and forage, five hundred under grass for hay, and three hundred under peas. Seventeen years after Grigg first set foot on the property, a village surrounded the homestead with 150 full-time staff, including shepherds, dairymen, teamsters and other stockmen, fencers, gardeners, general farmhands, as well as bakers, blacksmiths, carpenters, saddlers, butchers and bacon-curers, and those involved in boiling down surplus sheep carcasses into tallow.

Grigg convened the meeting in November 1881 that formed the Canterbury Frozen Meat Company and became its first chairman of directors, a post he held for nearly twenty years. In one season during the 1890s he bought 80,000 sheep and lambs for fattening and exporting to London. In one day he drafted 4400 lambs on Longbeach for the freezing works and the average carcass weight was over forty-two pounds. In 1894, 5500 acres were under wheat.

The leader of this extraordinary community, paying an estimated £8000 a year in wages, was a fair-haired man of average height, limping slightly on a leg damaged in a childhood accident, his face covered by a full beard, always on the move either on foot or horseback. Grigg's organisational and administrative talents were almost as legendary as the foresight that enabled him so often to predict trends and market developments. His reputation was for unalloyed honesty and remarkable generosity to his loyal staff.

His obituary in the *Press*, Christchurch, 1901, said Longbeach:

> [H]as been described by a widely-travelled expert as the best farm in the world. It has long been one of the show places of the colony, regarded by the people of Canterbury with pardonable pride ...

A sod hut in Canterbury of the sort Grigg would have lived in while he broke in his Longbeach property.

Zealand sheep. The breed was brought in from Australia in large numbers from 1840 to serve as the basic flock sheep, the role it had played in Australia. The merino pickety-picked in its goat-like way at pastures lush and sparse over the hills and valleys of the whole country for more than forty years and the woollen industry had been woven from its silken fleece. But in another forty years it would represent only five per cent of the national flock, almost exclusively in the high country of the South Island. Not that the solution of cross-breeding was new. It was simply accelerated. Some sheep farmers had been putting Southdown or Leicester rams over merino ewes since the earliest days to produce a less skittish sheep with a better meat carcass for the urban market – a cross that would do better on the stronger, comparatively rank English grasses. The merino's favourite food was the fine, feathery native pastures.

During the late eighteenth century British agriculture had gone through a revolution as enlightened farmers concentrated on breeding and on pasture development as a response to the country's burgeoning population. Although the enclosing of common land had socially distressing consequences for village peasantry, it did allow firm control of stock and led to selective breeding towards higher productivity of sheep and cattle. Fenced paddocks also spread the principle of rotating crops rather than having to leave land fallow every few years to prevent its exhaustion.

In the 1760s Englishman Robert Bakewell became the first to adopt some of the animal breeding methods still used today to enhance inherited qualities he wanted to perpetuate in his animals. Not only were his breeding techniques innovative, his concept of breeding sheep and cattle for meat was new too. He helped create the 'roast beef of

The Border Leicester was one of a number of British breeds used in cross-breeding to get top-flight carcasses for the frozen export trade.

England' myth because, before his time, the main roles in life for cattle were as milch cows or draught animals, and sheep were most valued for their wool not their meat. Bakewell changed the scrawny commons-grazing sheep of his county into the English Leicester, famed for its delicious mutton.

Other farmers in Britain followed his example and bred up their own local stock. This breeding revolution that began in the eighteenth century expanded in the nineteenth, and had important ramifications for pioneering New Zealand farmers, who were nothing if not innovative. A number had imported the Leicester, its spin-off the Border Leicester, and also the Lincoln, Romney Marsh, Cotswold and Southdown. Farmers were trying for sheep better adapted to heavy country and the new breeds and cross-breeds produced better fleeces of their long-stapled wool than the same breeds produced in Britain. As early as 1874 a Christchurch *Press* report announced that Canterbury farmers had been cross-breeding so extensively over the previous two years that the crosses outnumbered merinos at the saleyards. The tempo then sped up, driven by the demand for meat for the export trade. By 1892 there were only six million merinos in the national flock of eighteen million.

Over the following decade the volume of mutton in the London export trade diminished as the crossbred fat lamb took over. Partly this was in response to a marketing difficulty. South American countries were dropping huge quantities of high-quality mutton into London but they couldn't match the plump and succulent New Zealand lamb which appealed to the English palate.

The heavyset Lincolns, one of the first breeds fixed during the agricultural revolution in England, arrived in 1840 but the most influential importation was in 1862 by the New Zealand and Australian Land Company. It was used immediately over the merino to produce successive generations of crossbreds and what became known as the 'colonial half-bred'. Lincolns were the most common sheep in the North Island until the turn of the twentieth century. The first Romneys were put on pasture at Waiwhetu, in the Hutt Valley, by Alfred Ludlam, in 1852. An English Leicester flock was established in Auckland in 1853 and another in Canterbury in 1865. Border Leicesters were brought in by the New Zealand and Australian Land Company and by Mr George Murray of Taieri in 1859. A Scottish shepherd, James Little, arrived in Otago in 1863 with twenty-two Romney Marsh ewes and nine rams for the Corriedale estate of Dr George M. Webster. Little decided the tussock grass of the station did not entirely suit the Romney and he also aimed to develop a dual-purpose animal more suited to New Zealand conditions than either the merino or the long-woolled English breeds, so he set about producing an inbred half-breed, using Romney and Lincoln rams over merinos, and he called them Corriedales. They were the first New Zealand fixed breed.

The same William Davidson whose foresight and organising acumen later launched the frozen meat trade was farming on behalf of the New Zealand and Australian Land Company at The Levels in South Canterbury in the early 1870s. He studied the breeding techniques of Little and fixed a Corriedale flock there. It was a shrewd adaptation that solved some of the problems with merinos caused by climate and environment. Corriedales and some unfixed crosses with the heavier English rams over merino ewes produced better carcasses (albeit coarser wool) from the English grasses that were then beginning to carpet New Zealand's grazing areas.

But the breed that came to dominate all the others was the Romney. A fastidious mother and good milker, the breeding ewes produced useful strong wool and were ideal for producing fat lambs by Down rams. This cross became the solid economic base of the meat trade with Britain early last century. By the time the merino had slumped to five per cent of the national flock, in 1921, the Romney outnumbered all the other breeds put together. By World War Two cross-breds made up more than seventy per cent of the national flock with Romney about fourteen per cent, merinos and Corriedales around five per cent and all the others less than one per cent. The Romney remained dominant. It had been adapted to New Zealand conditions from the original English Romney Marsh importations. However, declining fertility spread through the national flock in the post-war years, even though scientists helped arrest it by convincing farmers to cull infertile ewes as young as

possible through a national breeding programme called Sheeplan.

Probably the major reason for the decline in Romney fertility was the breed society's concentration on conformation and other physical qualities not directly related to actual breed performance and productivity. New Zealand Romneys became narrow in the pelvis with wool on the face and legs, whereas the indications were that a clean face and legs were genetically linked with fertility. Veteran gun shearer and machine-gun publicist Godfrey Bowen, writing about the evolution of shearing techniques in his book *Wool Away*, said: 'The present-day Romney, with wool to the toe and even on the top lip, is much different from the clear, open-pointed Romney of fifty years ago.' The conformation and performance of the breed was dominated by stud breeders, mostly concentrated in the Manawatu-Rangitikei region and, even though they had fixed the specifications of the Romney, problems affected their flocks too. Low fertility, still an issue especially in the North Island, was a reason for the development of two new breeds, Perendale and Coopworth, both fixed from Romney crosses, which both grew rapidly in popularity. The breeding-up of fertile strains within the Romney flock through Sheeplan, and through the efforts of individual farmers who had become

The Cheviot Hills Station homestead of Dr George M. Webster (left) demonstrates the affluence of the runholders before the big estates were broken up by the Liberal Government. It was on Cheviot Hills Station that a Scottish shepherd, James Little, developed the first New Zealand fixed breed, the Corriedale, using Romney and Lincoln rams over merinos. When the estate was sold in 1893, hundreds of people arrived to inspect and, in some cases, buy Corriedales for their own properties.

more conscious of the need to select for fertility, enabled the breed to hold its premier position in the national flock.

The character of the wool trade was completely changed by the eclipse of the merino. New Zealand's major output became coarser wool, much of it for the carpet trade – which grew vigorously both overseas and here from the early 1960s. The Romney fleece was valued as the major component of woollen carpets, desirably mixed with a coarser wool such as from Scottish blackface which was no longer available in New Zealand. Yet by the time the need for its coarse fleece for carpets grew, the disease risk had put up a barrier against its reimportation. But the industry was reprieved from the continuing reliance on imported blackface wool by Dr F. W. Dry, a research scientist at Massey Agricultural College who developed an adequate replacement, the Drysdale breed, more or less by accident, in the early 1960s.

Refrigerated shipping had an indirect but positive effect on arable farming practice during the 1880s and 1890s. Among the by-products of freezing works were fertilisers, and this encouraged farmers to sow turnips and rape with fertiliser either before or after a cereal crop. This rotation with the additional cultivation enhanced the fertility of the land, improving wheat and oats crops, and made food available for fattening stock. Thus a mixed farm could operate independently, or the farmer could take surplus store stock from hill-country properties after shearing and fatten them for the works. This practice was supplanted in the first two decades of last century by spring topdressings of superphosphate on both cereal crops and pasture.

Negative characteristics become positive

The Drysdale sheep has a gross, hairy, long-stapled fleece, lacking in crimp – the sort of wool Dr F. W. Dry and his team at Massey Agricultural College were attempting to eliminate from the Romney – until they realised it was just what the carpet industry wanted.

In the 1920s wool processors were reporting what was considered an undesirable fault in New Zealand's cross-bred clip – too many hairy fibres in the fleece. Back in 1932 the Governor-General, Lord Bledisloe, in a lecture on agricultural research, said: 'Kemp or hair in the fleece, especially of Romney Marsh sheep, is a notable and growing defect in the wool of New Zealand flocks and much research both at Massey College and in England has been directed towards its elimination.'

Dry was a foundation lecturer at Massey, and because he had worked in England on the inheritance of fleece colour in Wensleydale sheep, and also on the coats of mice, he was asked to research how what they called halo-hairs were inherited and how they could be bred out. By 1940 he had found that the culprit was an N-gene, and in the process fixed the eponymous Drysdale, a breed in which this gene was dominant and whose wool was similar to that of Scottish blackface and some Asian and Middle Eastern breeds not present in this country but valued by the carpet industry. A Christchurch-based carpet manufacturer had applied without success to bring in Scottish blackface sheep, so the heavy wool had to be imported for blending with other yarn in New Zealand carpets.

The Drysdale produced slightly more wool on average than the Romney, had a similar lambing percentage, was slightly longer in the leg and higher at the shoulder, and was open-faced. Rams were easily recognisable because of horn growth. The commercial development of the breed began in 1961 when control of Drysdale wool production was vested in two companies which bought rams from accredited breeders and leased them at an annual fee to producers of fleece for the carpet industry.

The Drysdale was a useful animal for a few decades but went out of fashion with woollen carpets – and no other reason justified the breed's use, especially because it needed to be shorn twice a year, a costly business. Flocks have been largely disbanded.

Scottish blackface had been brought into New Zealand in early years but had disappeared from the national flock. The breed was not the only one known to have been imported and then to have faded out. Among others were Norfolk and Oxford Down breeds (now brought back), the Kerry Hill, Cotswold, Wensleydale, Roscommon, Dartmoor and the Tunis. No doubt some of these are still represented in the gene pool in this country. The two New Zealand breeds adapted from the country's varied stock sources were the Corriedale and the South Suffolk.

Dairying was the boom industry in those early years of last century. Because milk needed more careful handling and more complicated processing than meat, dairy exporting developed more slowly. But, although many settlers fattened lambs on early pastures after they had wrested their farms from the bush, dairying could better support under-capitalised smaller holdings by providing a more consistent cash flow. Wool and meat prices were generally subject to wider fluctuations of price. A result was that milking cows were put on a lot of land only marginally suitable. Those who made it through to a high enough level of equity in their property aggregated their neighbours' properties and moved into sheep; or perhaps got onto better land as dairying gradually consolidated in the most prolific grass-raising areas where soil fertility was adequate and warm rain fell throughout the year. In this way, much of the land originally broken in for dairying was later turned over to sheep.

The development of refrigeration was a massive boon to New Zealand farming. Wool was the early export winner, capable of being shipped anywhere in the world without the worry of deterioration, but meat and milk have become the commodities that enabled New Zealand to escape Third World status. Also, freezing technology has been refined over the years, providing sophisticated chilling systems that keep produce very close to fresh. Allied with fast and efficient transport using containerisation, chilling has enabled the export of meat cuts to the top end of international markets and also the delivery of fish and a range of vegetables and fruit in near pristine condition to consumers right around the world.

Merino lambs in a yard on Wadworth high-country farm, Marlborough. The merino was the first major breed of sheep in New Zealand but as other meat and wool breeds were developed that were more suitable for this country, its numbers fell steeply. However, a niche market for very fine wool at good prices has kept the merino in business, mainly in the high country of the South Island.

CHAPTER NINE

EACH DAY WAS 'JUST SO MANY TREES'

As new settlers made their way in bush country, waiting for a burn to end or for pasture to grow, they had assets but little or no income. Many took jobs laying tracks if a railways project was nearby, or on road-making teams.

THE DESTRUCTION of the North Island forest was a sad and indiscriminate land rape; or, as Professor G. Jobberns of Canterbury University once said, it was 'the outstanding achievement of our people in the making of the present grassland landscape'. Perhaps it was both of these things.

The first European visitors to this country were awed by the statuesque giants in the great rainforests – as commercially impressed as aesthetically. The kauri, often with twenty-metres-plus of clean trunk to the lowest branches, was coveted by shipbuilders for masts and spars, as well as deck and other planking. The Royal Navy got involved early in the trade which grew quickly in the early years of the nineteenth century. Kauri grew only in the north of the North Island and the depredations went deep into the hinterland of Northland and the Coromandel Peninsula. The Northland forests were ravaged also by out-of-control fires set by gum-diggers and by both Maori and Pakeha farmers involved in cropping to victual ships and to feed the embryonic towns on the northern coasts. With casual profligacy they often destroyed thousands of hectares to clear a relatively small arable area. But the final destruction of the forest came later from the systematic assault by settlers who had bought bush and needed farmland.

On one hand, successive governments had conservationists fulminating from the early years of Parliament. A report compiled by Sir James Hector, Director of the Geological Survey, from information supplied by provincial superintendents and released in 1874, was alarmist in its assertion that about half the forest area of the North Island had been destroyed by then. Even if this included large areas of scrub and wasteland which may not have been in heavy bush when the Europeans arrived, it was still an arresting claim. On the other hand, the government was beginning to pour millions of pounds into roads and railways to take settlers into the bush where their survival depended on their ability to tear down trees and get the land into grass. Such people left conservation worries to those who could afford them.

An anonymous poet waxed wordily about 'this desert of logs this dead disconsolate ocean' and:

> The prone logs never arise,
> The erect ones never grow green,
> Leaves never rustle, the birds went away
> with the Bush . . .

And prose writers prattled, like a woman quoted in *Brave Days* describing a 'bush paradise':

In the nineteenth and early twentieth centuries, the timber industry required hard work by tireless men using two-handled crosscut saws. Logging and burning valuable indigenous trees had caused deforestation so dramatic that alarm bells were ringing among local governments as early as the 1870s.

Top: Bishop Ditlev Monrad, disgraced in his homeland of Denmark because he was Prime Minister when his country was defeated by Bismarck's Prussian army in a war over Schleswig-Holstein, came to New Zealand with his family in 1866 and settled at Karere, now Longburn, in the Manawatu. He was one of a number of Scandinavian immigrants who came to wrest farms from the stretch of bush in the Manawatu and southern Hawke's Bay region. After three years the bishop returned to Denmark and steered clear of politics but two of his sons stayed on.

Above: Eileen Duggan, published in both Britain and the United States, was the first New Zealand poet to achieve some international recognition. Much of her work was nature verse and reflected rural life.

There was a great demand for land [in the eighties] and cheap areas from five to two thousand acres were offered by the Manawatu Company. This offer attracted the attention of a widow in Canterbury, who, for eight hard years, had worked at teaching dancing, painting, and many other things in order to educate her three children. The eldest girl was then teaching and needed no help, one child was at school in England, one in Melbourne, refrigeration was operating – her instinct was right – it was land she must have . . .

The eighteen-year-old daughter couldn't resist such enthusiasm, and, like babes in the wood, they left for the North Island. In Wellington, instead of a wicked uncle they found very good friends to give advice and guidance. The Atkinson Government had prepared and was ready to open the first Village Settlement (Levin). Our selectors were strongly advised to consider this block . . . and were taken in the Public Works' train to see it. Paradise complete, without flaw or serpent, unfolded itself before their astonished gaze. It was a block of solid native bush in all its original glory. Very few New Zealanders have seen such a sight . . . Tall milling timber grew close and straight like a field of giant wheat-stalks. And the undergrowth – the incredible undergrowth! Nature had filled every available inch with something exquisite. Huge tree ferns found space overhead among a riot of trailing vines. Beneath, smaller ferns pressed for space, shoulder high, and underfoot was carpeted thick with miniature ferns and moss, of magic liveliness . . . A gang of Austrians were at work. They wore coloured handkerchiefs round their heads and called to one another in their musical foreign voices that mingled with the bellbirds and tuis like a rhapsody. The daughter was faint and dazed and overwrought with too large a draught of beauty . . .

This description was written from the candyland of hindsight, with the graft of gaining grass glossed over with all its unremembered pains: 'Well, it became necessary, after a while, to farm fairyland, and that was hard and sordid, but the daughter found romance there, and afterwards true happiness, and the mother a generous measure of prosperity. And so all was merry.'

Most of the voices from the age of the bush-breakers belonged to dilettantes whose sweat had never broken in the swelter of the summer as a saw bit into the time-toughened sinews of the big trees. The bushmen had no bard, unlike the bush itself, unlike the squatter, the drover, the swagger. They were less well educated, and had less time than the southerners who left voluminous diaries and letters and even published books. They weren't as literate. Many came from Scandinavia and Yugoslavia and had no English. Later, they were celebrated in Eileen Duggan's 'The Ballad of the Bushman':

Only a fool would eat his heart out so,
And I am old enough to have more sense,
But lying by the window makes it worse –
Look out across that fence!

I have to die, they say, so let it come.
A clean break is the best I've always said.
And better burn than blight for any crop.
So I am better dead.

My wife says do no fretting over her.
She will be soon upon my heels up there.
We have been in it ankle-deep at times.
She always pulled her share.

No son of hers was ever rocked to sleep.
She bore me three here in the bush alone.
She was as hard as maire in some things,
But faithful to the bone.

Each day to us was just so many trees.
Night was a minute then. Short is the word!
Ah who will wake me out of my last sleep?
Her clock was the first bird.

It is not for my sons I grudge to go.
Good lads like them will never come to harm,
The neighbours envy us the name they have,
But God, if one of them could farm!

Settlers breaking in bush country put up with hardships hard to imagine today. The first home (above, this one in the Manawatu) almost always involved calico or canvas for shelter, an open fire for cooking and outdoor toilet facilities. A second home (left) was usually small for the big families of the time and made from pit-sawn timber, but was a palace in comparison.

One thing is sure: for most bushmen each day was 'just so many trees' in a harsh regimen of unremitting work. As G. C. Petersen wrote in his essay *Pioneering the North Island Bush*:

> It is difficult to realise in these days the price the pioneer bush settlers had to pay to gain their small and often uneconomic holdings. The inner history of the bush settlements is one of hardship, isolation and privation unknown to the settlers of the open country, of men crushed by falling trees, drowned in swollen bush streams, and of men, women and children lost in the limitless bush and sometimes perishing from starvation or exhaustion. Children died because their parents could not afford the cost of bringing a doctor to their isolated homes, and many a desperate man or woman carried an ailing child ten miles and more along muddy bush trails in search of medical help. The crowning disappointment for many of the Scandinavian settlers [in the Forty-Mile Bush] was the discovery that the forty-acre farms allotted to them were, when cleared and grassed, hopelessly inadequate to support a family . . .

The last of the accessible bush held out until well into the twentieth century. In fact, the pioneering phase of New Zealand farming did not come to an end until just after World War Two. In the 1950s and 1960s, when I was working in the country and later writing feature articles for the *Journal of Agriculture* and the *Weekly News*, I spent much of my time driving through countryside. Most of the smaller farms were still set in stark landscapes – a house with a small, rough-dug vegetable garden, some macrocarpa windbreaks, a few outbuildings, and a house paddock fenced in by sagging wire. Gradually, after enough years of good returns and diminishing debt, farmers and their wives planted pretty trees, flowers, built white-painted wooden-railing fences around house paddocks and then new homes. The land at last looked not only lived in but loved. But the wan bones of old forests still haunted the hillsides of some remote areas where farmers gave up or just hung on.

Nothing is as indiscriminate as fire which, once it took flight, was often impossible to control. Pictured is the start of a blaze that razed the bush and scrub on 30,000 acres on the Puketoro Station, north-west of Tokomaru Bay, circa 1900.

Top: It was sometimes years before rough, burnt-over country was able to be ploughed and large teams of bullocks were required to pull the single-tine implement.

Right: The last task in clearing the land of bush was to remove the stumps, like this huge rimu stump which took a team of men to prise from its hole near Levin, in 1899.

Some areas had been so thoroughly denuded even by 1897 that the Secretary for Agriculture, Mr John D. Ritchie, regretted on Arbor Day that year that the interest in the planting of trees was not being kept up. He wrote, 'There are many odd corners about the farm where a few trees might be put in. The want of shelter for stock is much felt in exposed districts, and the great necessity for it is being forced upon stockowners. The indiscriminate clearing of the bush and ploughing of the paddocks has removed the shelter which formerly existed; therefore it is imperative some substitute should be provided.'

Refrigerated shipping created export meat and dairy industries based, especially in the case of dairying, on the intensive farming of smallholdings. The Liberal Government of the 1890s repurchased large holdings for subdivision and in other ways determined to provide land for intensive farming by people who would work it despite limited capital. The main regions where forest gave way to farm, with forty or fifty or more acres allocated to each family, were Taranaki, Manawatu-Rangitikei, the Seventy-Mile Bush that marched from the Wairarapa into Hawke's Bay (soon to become the Forty-Mile Bush), South Auckland and parts of the Waikato and, later on, the King Country. Most of it was rolling to steep. The settlers, although better equipped with iron and steel tools, used land-clearing techniques similar to those of Maori in earlier times. The first step was to choose the right sort of bush, not only for ease of clearance but also for potential fertility. Rimu was to be avoided, and while white pine and pukatea suggested rich soil they were hard to burn. Tawa land was regarded as the best.

was set up because its principals believed Southland was more suited to dairy farming than grain-growing or the breeding or fattening of stock. Local farmers, however, did not seem to agree, Brydone said later, so the company had to buy 350 milch cows of its own for supply. That was in 1880. They made only cheese at first, which they sold for good prices within New Zealand and Australia, and later, less successfully, in London.

Once local farmers, persuaded by results, recognised the benefits and profitability of dairying, the company leased them farms from Edendale Estate land and sold them cows at a price that enabled them to get started despite their limited capital. Soon these farmers were supplying all the milk required by the factory. Overtones of social altruism accompanied the project. Brydone said later the company paid slightly more for milk than factories elsewhere because, 'We prefer . . . to get our rents paid regularly, and see the farmers succeeding, to getting big interest, or even any interest at all, on the capital in the factory.' In 1890 Brydone told a parliamentary select committee the company had 'the very best appliances' for making butter but could not find enough butter-making experts 'even if we have men who seem to take every care and who are as good, I believe, as it is possible to get in the colonies'. This concern about expertise at factory level persisted for years.

Several more factories capable of manufacturing both butter and cheese were set up in the early 1880s. Canterbury, Otago and Southland were still the best-organised farming centres and dairying attracted interest; but in the North Island, as tensions with Maori subsided and appropriated Maori land was allocated to aspiring farmers, the bush was being rapidly razed and dairying was starting to explode in the areas where it has flourished most ever since – mainly in Taranaki, the Waikato, the Bay of Plenty and Northland, but also in Manawatu and the Wairarapa.

By 1885 butter exports totalled 24,923 hundredweight worth £102,387 and cheese 15,245 hundredweight (£35,742). Twelve years later, after the spread of the centrifugal separator (introduced in 1880), and the more general use of refrigerated shipping, butter exports were up to 75,287 hundredweight (£297,515) and cheese to 71,662 hundredweight (£135,711). During those twelve years the understanding grew that dairying would need to become a factory-based manufacturing industry to gain a consistent quality of product for the required volume, especially with butter. As dairy produce moved into the yawning British market greater evenness of quality was essential.

The 1890s was a watershed decade for dairying. A select committee of the House of Representatives discussed and reported on the industry in September 1890. It noted the value of exports of butter had increased during the decade from £8350 to £146,840, and of cheese from £1983 to £67,105. Despite the optimism these figures

The early dairy factories needed to be close enough for farmers to get there and back from home by horse and cart every day. Taranaki developed earliest as a dairying province and, as this 1903 picture at Stratford shows, farmers waited in long lines to deliver their milk. Foregathering and talking while they waited their turn added to the stridency of dairy farmers' political views, and if the sun was hot the milk in the cans 'turned' and led to 'fishiness', a common early complaint about New Zealand butter.

Opposite: Equipment that was essential for the farm dairy (clockwise from top left): a piece of equipment used for cutting and embossing pats of butter; a traditional milk can; milk ladles of varying volumes, such as a pint and a quart; and a hand-operated butter churn.

Cheese-making required more expertise than butter-making, and in the earliest days some households did not bother with the task even though the most common cheese was the simple cheddar type. In the early cheese factories, labour was required to help with most stages of the process, such as draining off the whey, which is what the cheese-makers are doing here, about 1910.

inspired, the committee considered the export trade was in an unsatisfactory state, that butter was 'frequently made in a faulty manner', and that even good butter was often spoiled in transit 'through defective carrying and transhipment arrangements or the use of unsuitable packages'. The many experts who appeared before the committee gave conflicting opinions on the cause of both butter and cheese arriving in London in poor condition – including on whether or not it was made badly here through ignorance; whether freezing did or did not damage butter; whether tawa kegs tainted produce and totara did not; whether shipping delays caused produce to be left waiting in bad storage conditions; and whether shipping companies provided space suitable for dairy produce as they did for meat and yet charged more for it. In its report, the select committee recommended the government should give every attention to developing dairying because it believed the market in England to be 'practically unlimited' and that European countries had nearly reached the limit of their production.

Inspectors were appointed in London to check on quality and experts were brought to New Zealand from Denmark and Canada. An experienced Scottish dairy instructor and organiser John Sawers, appointed Chief Dairy Expert by the government the year before the select committee hearings, was already having a heavy impact on the industry. His first job was to give practical demonstrations and lectures on cheese- and butter-making at the South Seas Exhibition in Dunedin. He said the factory system was essential for the future, advocated a co-operative system, and offered his conclusion that the industry had problems that sprang from 'the want of knowledge of many of those engaged in the manufacture and from the want of cooperation in the interests of the factory on the part of the settlers and milk-suppliers'. A blunt man, he made enemies in the industry but was so effective he became the first Chief Dairy Instructor when the Department of Agriculture was formed in 1892. It was later said of him that his energy and imagination were instrumental in the evolution of a well-organised co-operative factory system.

By 1890 there were about 150 dairy factories in the country, only a shade over forty per cent of them co-operatives. But, as Sawers later pointed out, problems of quality control persisted despite government intervention with controlling legislation in 1892 and again in 1894. A government dairy produce grader, Mr A. A. Thornton, reported to Parliament in 1897 that it was 'a great pity' the Dairy Industries Act 1894 did not empower inspectors to prevent the shipment of the 'most inferior quality' of butter. He described one consignment of butter arriving in London as 'a perfect disgrace', and a consignment of cheese as 'weak in body and texture and defective in flavour'.

The first Taranaki dairy factories, both privately owned and co-operatives, had opened during the 1880s, and a key entrepreneur at the beginning was a Chinese storekeeper, Chew Chong, who had migrated to New Zealand during the goldrush period. In the Waikato, the butter pioneer was a Cornishman, Henry Reynolds, who established the Anchor brand with the first factory in that region. Reynolds expanded within the Waikato and then into Taranaki, but, like Chew Chong, was swallowed up by the co-operative movement during the 1890s. His company eventually became the huge New Zealand Co-operative Dairy Company, which continued with Anchor butter.

The co-operative movement helped shape a generally collective attitude towards export marketing. Although Chew Chong's exports of 'Taranaki wool', the white fungus from the mahoe trees, exceeded exports of butter and cheese during the 1880s, the overseas market for dairy products gradually expanded and was worth £197,000 by 1887, and £350,000 just six years later. During the period to World War One, individual dairy companies did their own selling through agents in Tooley Street, London's butter market. The industry was moving towards a fully collective marketing organisation when war broke out and the British and New Zealand governments came to an agreement that developed into the purchase of all New Zealand exports as part of the war effort. Demand meant cheese production grew quickly because it was such a useful wartime food.

The bulk-buying agreement ended in 1921 when producers went back to competitive selling – but only for long enough for the industry to realise this method dragged prices down. A group of seventy co-operative factories formed a limited liability company to market their produce jointly in London. By 1926 the company had ninety members. In the meantime, the government established the New Zealand Dairy Produce Export Control Board (progenitor of the New Zealand Dairy Board) and in 1926 it took control of exporting, including prices – only to run into stout opposition from free-traders in Britain. So the price-setting had to end but the concept of orderly marketing by co-operatives banding together remained. The industry arranged freight and insurance contracts, regulated shipping to ensure continuity of supply, and introduced a national brand, Fernleaf. Two New Zealand companies – Amalgamated Dairies Ltd and Empire Dairies Ltd – acted as wholesalers on Tooley Street, but in 1936 a new Labour Government installed a guaranteed pricing scheme by which it bought all the butter and cheese from producers at a fixed price. The plan was to smooth out returns to farmers but, although it lasted for twenty years, the industry was never happy with the pricing mechanism. A provision existed for government to support the industry fund when it went into debit but it was almost always in credit – with the result that when subsidies were abruptly tugged from underneath the industry in the mid-1980s, dairy farmers landed more softly than otherwise would have been the case.

In the depths of the Depression, the industry was in a graver crisis than at any time before or since. A survey in 1934 revealed that about half of New Zealand's dairy farmers were technically insolvent. A national scheme to help them refinance mortgages on a long-term basis at low interest staved off total failure until a pick-up in the world economy and dramatic intervention by the new government in 1936 resulted in a gradual recovery. During World War Two, Britain again became the sole buyer of all surplus dairy production in New Zealand, an arrangement that lasted until 1954.

In the early days of the dairy industry, small isolated herds in the North Island made transport to factories difficult. An answer was a proliferation of district creameries or skimming stations to which farmers could take their milk, have it separated by machines, take their skim milk and return home. The creameries delivered the cream to the nearest factory. However, mitigating against quality

One of the most inventive and enterprising dairying pioneers was a Chinese former peddler, Chew Chong, who became a storekeeper and opened a dairy factory in Taranaki in the 1880s. He helped dairy farmers increase their income by getting them to collect from their farms a fungus that was prized in China, and which he bought and exported.

Chew Chong, dairying pioneer

Chew Chong was among a number of Chinese who came to New Zealand during the Otago goldrush. He had no luck as a prospector and moved to Taranaki in 1870 where he became a peddler of household goods, trinkets and children's toys. While he was walking through the province's recently cleared bush, selling his wares from door to door, he recognised a white fungus, commonly known as 'Jew's ear', which was a delicacy in China. The fungus flourished on the dead trunks of tawa, pukatea and especially mahoe trees – or 'whiteywood' as the mahoe was called.

He started an export business to China in what became known as 'Taranaki wool', and earned more in overseas funds in the early years than dairy products – an estimated £300,000 over thirty years. Chew Chong bought Taranaki wool for fourpence a pound – the same price butter was fetching locally. For hard-up settlers this was a boon. Their wives and children spent hours gathering the fungus in the forest.

He set up shop in New Plymouth and then began buying butter for export to Australia while also importing Chinese goods for sale. In 1884 he relocated his shop to Eltham, where he again bought butter, or took it as barter for goods in his shop. With refrigerated shipping available, he audaciously exported consignments to London, but without financial success because of the condition in which it arrived. Nevertheless, he built a butter factory in Eltham in 1887, hired an expert butter-maker and set up a network of skimming stations. Two years later he won a silver cup at Dunedin's South Seas Exhibition for the best half-ton of creamery butter packed and ready for export.

Chew Chong was a genuine pioneer and entrepreneur. His was the first butter factory to install refrigeration equipment, and the first to make use of a stamped butter brand (the Jubilee brand in honour of Queen Victoria). He invented a butter-working machine and introduced a system on his farms that resembled share-milking. However, he was eventually swamped by the burgeoning dairy co-operative movement and gradually lost his suppliers.

Chew Chong came to be admired and respected in his adopted province – and that was no mean feat in the days when the 'Yellow Peril' was the equivalent of a later era's 'Reds Under the Bed'.

The spread of the centrifugal separator meant farmers need take only their cream and not the full milk to the district dairy factory each day.

Following pages: Cattle on a Ngatea dairy farm eating palm kernels. An imported by-product of the palm oil industry, the kernels are cheap and sometimes used as a supplement to maize silage or grain feed.

produce were the idiosyncrasies of individual farmers. The standard of hygiene was not always high, either, in cowsheds where milk was stored – sometimes for two days in the summer heat at the height of the season. For many years, the chief complaint against New Zealand butter in Britain was its 'fishiness', a smell and taste that resulted from cream that had gone off before the butter was made. The creameries were, nevertheless, a necessary convenience for remote farmers, some of whom were so isolated by bad roads, hills and streams that they had to transport their milk in cans on pack-horses.

The spread of the home separator and improvements to roads brought about the demise of the skimming stations. With the reduced burden of only his cream rather than full milk, each farmer began his day with a visit to the district dairy factory – his dairy factory. After a slow start, the concept of the co-operative dairy company spread like a faith – an extension of the smallholder's desire for as tight a mastery as possible over his destiny; and, of course, it gave him a higher return from the product his raw material was processed into.

Whereas the meat industry in most areas was dominated by relatively few runholders who were powerful enough to buy shares in private freezing companies if they wanted to, the dairy supplier was almost invariably a smallholder with limited capital. (Two-thirds of a total of 929 holdings in Taranaki County in 1881 were between ten and 200 acres.)

Dairy co-operatives were a thing of the times, and popular in other countries as well, including the United States. Although co-operatives in New Zealand were still outnumbered in 1900 by 152 proprietary factories to 111, the movement was unstoppable, and within a few years the proprietary dairy factory was virtually extinct. In 1902 the secretary of the newly formed Department of Industries and Commerce, T. E. Donne, advocated in his annual report that freezing works should also be run as co-operatives, saying: 'in some instances State control is strongly advocated. Whatever may be the merits of these different schemes there can be no doubt that the distribution and selling of frozen meat should be put on a better footing than it is at present.' Although some powerful meat company co-operatives did emerge, the ideal never manifested as anything near as strong a movement as it did among dairy farmers. As exports grew from £208,000 in 1890 to £1 million in 1900, British importing houses began to provide loans for capital and felt more secure where the suppliers were involved. However, the importance of privately owned factories in the early days, especially in Taranaki and the Waikato, shouldn't be undervalued.

As roads improved – and especially after tanker collection spread throughout the country – factories tended to become centralised and bigger. By the end of World War One dairy factories around the country numbered 638 and cows totalled about 750,000. Cow numbers continued to rise to about two million in 1960, by which time manufacturing plants had dwindled to 268. Then the rush of amalgamations began with the aim of gaining economies of scale. During the following twenty years, the tanker became ubiquitous, picking up wholemilk from even the most distant farms and delivering to bigger and bigger and more central factories. As herd sizes climbed from an average of sixty-five cows to 122, amalgamations further reduced factories to 105 by 1980. Output per factory more than doubled. As the factories got bigger, the standards of hygiene and quality control also improved. Butter-making in particular became mechanised, with the cream poured in one end of the operation and packages popping out the other. Cheese-making became similarly mechanised where the output of a single type was large enough in volume. Dairying developed over the first sixty years of the twentieth century as a compact industry with factories centred on intensively farmed grassland of

Breeding up the herd

Milk always demanded more care and scientific knowledge than any other bulk primary product as it journeyed from the milking shed, through factory processing, to international markets. Wool was easy: you clipped it and packed it. Meat required easily achievable expert butchering and temperature-controlled storage. The quality of these products depended mainly on experienced husbandry. Dairy farmers needed similar husbandry skills, but milk is a fragile and perishable product requiring careful attention on a daily basis by thousands of small farmers and processing workers.

For more than twenty years after butter and cheese became export commodities the industry struggled to achieve a consistent level of quality. Inadequate facilities and underskilled people working on many of the hundreds of undercapitalised farms and factories meant assistance and regulation by central government and industry organisations were pivotal to building an export trade. Twenty-five years after refrigerated shipping opened up international markets, the *New Zealand Official Yearbook* reported: 'The government still spends large sums of money in teaching the art of butter and cheese making, the proper method of grading, packing and shipping.'

Gradually, dairy advisers at the farm and factory level in New Zealand raised the level of equipment and expertise and inspectors in Britain reported on the quality of product on arrival. The 1907 *Yearbook* also said the government recognised the importance of breeding 'better strains of cattle for the production of milk of superior quality as well as quantity' and thus had imported some first-class sires for dairy farmers not in a position to secure the best bulls: 'The service of these bulls has been fixed at a nominal scale and is only available for selected animals.' So the government was involved in breeding up the national herd at least two years before the first testing of factory suppliers – to the Dalefield Dairy Company, near Carterton. Eight hundred and fifteen Dalefield cows were tested for butterfat yield in 1909.

The desire to raise production nationally by identifying high-production cows spread. In 1923 a group herd-testing scheme with a

Preston's Famous Cow Detectors was on the road in 1922, trying to persuade dairy farmers of the virtue of its 'herd testing group system'. By 1935, herd testing had been called 'a factor of incalculable value in the life of the industry'.

staff testing officer began in the Waikato with 7500 cows. The government and industry realised the value of group schemes, became deeply involved, and by the end of the 1920s the New Zealand Dairy Produce Control Board, progenitor of the New Zealand Dairy Board, reported that increased butterfat production per cow was testimony to the value of the herd-testing movement. Between 1922 and 1930, per cow production grew twenty-nine per cent, from 175 pounds to 225 pounds (approximately 79 to 102 kilograms).

By 1929 the twenty-seven herd-testing associations were amalgamated into six, each with a consulting officer to act as an advisory link between the associations and the Department of Agriculture. The scheme was funded jointly by the government and the dairy board. The government's support helped the movement gain traction. In 1935 E. J. Fawcett, Department of Agriculture farm economist (later Director-General of Agriculture), wrote that the systematic testing of dairy cattle on yield had become an integral part of management on many farms and rendered great service to the dairy industry by the highest quality, and with research, production and marketing co-ordinated by producer in their various incarnations, by the Department of Agriculture, by dominant manufacturing co-operatives, and by a number of research and regulatory bodies. Government-sponsored research was predominantly empirical and the progress was striking between the 1920s and the 1960s when Britain contemplated joining the European Common Market.

But as early as 1920 the importance of rich and sturdy permanent pasture types, and the need for lime and superphosphate, were fully understood. The best farmers were grazing their paddocks in rudimentary rotational systems, making silage for winter feed. There was more emphasis on fodder crops but the basic approach to pastoral farming was settled. Later developments were mainly refinements, many of them made possible by technological advances in farm machinery – electric fences, tractors, trucks – the whole range of appliances and machines made possible by the wider use of electricity and the internal combustion engine.

Perhaps the most influential man on pastoral farming in his time, Dr C. P. McMeekan, the

providing a measure by which any farmer was able to study his farm and herd management. The element of competition introduced through the press and agricultural publications had spurred farmers on to greater effort and improved management: 'Thus testing has become a factor of incalculable value in the life of the industry.'

Following more than a decade of research into insemination techniques by the Department of Agriculture's Dr John James, the first commercial artificial-breeding service began in the Waikato and Taranaki in 1949 in a bid to grade up the national herd. A year later the service was taken over by the Dairy Board. Artificial breeding proved so effective in herd improvement, especially as techniques improved dramatically, that it expanded rapidly. By the 1971 season, 716,000 cows – thirty-two per cent of the national milking herd – were tested and 961,000 – forty-three per cent – were artificially bred through the service. Consolidation, centralisation and expansion of the scheme continued through to 2001 when dairy company amalgamation led to control over testing and artificial breeding by the farmer-owned, Hamilton-based Livestock Improvement. By the early years of the twenty-first century around eighty per cent of cows were being herd tested and artificially inseminated, and around ninety-nine per cent of all dairy replacements were identified at birth and details recorded on Livestock Improvement's database.

The progressive development of dairy farming through to the end of the twentieth century depended on scientifically improved pasture; superphosphate topdressing; herd testing on a national basis to elevate the performance of sires and individual cows through artificial insemination; carefully monitored rotational grazing to gain maximum economic value from pasture; and heavy investment in scientific research on the manufacturing side. The trend was always towards larger herds feeding on grass for as long as possible through the year, on bigger properties with hay for winter supplementary feed and silage for use as the season faded.

Right: A technician intercepts semen from a bull at Livestock Improvement for use in artificial breeding programmes. Proven high-producing bulls are used to breed up the national dairy herd.

superintendent of the Ruakura Animal Research Station, wrote in 1955:

> Thirty years ago our industry was already a long way along the road. The turn of the century had seen its emergence from primitive conditions and methods. Refrigeration had opened up the markets of the world. The internal combustion engine had revolutionised transport, land development and milking methods . . . Thus the first twenty-five years of the century saw a remarkable tenfold increase in the exportable surplus of butter . . .

However, although many of the rituals changed, the basic practice of livestock farming didn't alter that much – it was more of the same done better. McMeekan also wrote that in the previous thirty years the major changes in techniques of dairy farm management and animal husbandry meant that:

> The concept of controlled grazing as the logical and most efficient way of using pasture has found its applications on every farm. The electric fence has been a powerful force in this direction by cheapening sub-division costs and increasing the efficiency of control. New types and strains of pasture plant have become available together with the necessary knowledge for their

Before butter-making was automated it was made in huge churns. The one at the left of the picture, which produced two-and-a-half tonnes of butter, is about to be emptied but first is being sampled by factory laboratory staff to ensure it complied with the legal limits of salt, moisture, and bacteriological content.

As roads and transport improved, horse-drawn carts were followed by trucks hauling milk cans to the factory, then milk tankers enabled larger, more distant dairy factories to replace the local plant.

Controlled grazing became the standard way of most efficiently using pasture to keep costs down, with perhaps feeding out of supplementary hay or silage to get the herd through unusually dry months or the winter when pasture growth was at its lowest.

establishment and maintenance for maximum yield. Topdressing with phosphates is universal, while lime, potash, nitrogen and trace elements are widely used where necessary with spectacular effects on productivity.

Special techniques for pasture renovation and improvement have been developed. In favourable localities, the menace of dry summers has been removed by spray irrigation. Hormone and other weed killers have banished the bug-bear of most noxious weeds. All these have combined to increase both total yield and total annual spread of grass growth. The concept of fodder transfer from periods of plenty for use during periods of scarcity through saved pasture and through hay and silage has become widely recognised as a key tool for high production. Making hay, silage or both is the pre-occupation of every efficient dairy farmer today in the late spring and early summer months.

On the breeding side, the principle of progeny testing as a basis to national herd improvement is universally accepted. Bulls of proven merit or their sons from proven cows head the herds of dairymen seeking progressive improvement in their cattle . . .

The scientists' word had spread effectively. In 1921 the average fat per cow was 174 pounds (approximately 79 kilograms), with about one million cows in about 50,000 herds averaging nineteen cows. Exactly fifty years later, just over twice as many cows (2.2 million) averaged 274 pounds (approximately 124 kilograms) of fat on fewer than half as many farms (21,900) in herds nearly five times as big (95 cows).

Average fat production had not climbed markedly in the thirty years after 1941 when it got above 260 pounds (approximately 118 kilograms) for the first time. It passed 300 pounds (approximately 136 kilograms) several times in the good years after 1965 but dropped down to around 260 pounds on one occasion too. As it was in 1921, the best-managed herds were far and away above the average.

Change during the first seventy years of the twentieth century seemed inexorable but was steady. Then Britain joined the European Economic Community and ten years later New Zealand deregulated her economy, opening up her borders to the full blast of market competition. The quickening of world trade abruptly ended the orderly pastorale of New Zealand dairy farming, the hard but satisfying life through the recurring tenor of the seasons.

CHAPTER ELEVEN

THE GREAT IDEAL – LAND OF ONE'S OWN

Wrenching land from the bush in the North Island was a heartbreaking business, especially for the great majority who had very limited capital. But thousands of families were prepared to endure the work of cutting out their forty or fifty acres for the great ideal of land of one's own. Most of the land was especially suited to dairy farming, but until two decades after refrigeration butter and cheese were seldom profitable.

The cow cocky's lot was not an easy one during the first fifty years from the 1880s when dairying as an industry first became a possibility. Over the years, thousands of men in their grey shirts, or raw-wool singlets, with brown speckled working trousers tucked into knee-high gumboots, bowed by the weight of mortgages, groped through their paddocks well before the sun showed, herding their cows and their wives and kids into the cowshed to start a day on which they would see the sun come up and go down again leaning against a cow's flank.

Milking was a family occupation. The farmer would be outside most of the day, watching the grass warily for the right degree of growth as he moved his herd round the paddocks, feeding out hay in the winter, turnips or silage from the late summer if he grew enough or could afford it, repairing fences and gates and machinery and milking systems, his face blotched with the sun, the skin on his forearms and the backs of his hands like potatoes baked in their jackets. Meanwhile, his wife would be inside the house between milkings, and if she had a large family, as many did, then her day was even longer and possibly harder. The poet Eileen Duggan spoke for the farmers' wives in 'The Milker':

She is up in the dark and out in the cowyard.
Sometimes a star is left in the air,
Through the cold interregnum between night and morning,
A timid vice-gerent, but what does she care?

Milking and stripping, milking and stripping,
She leans in against each cow as she stands.
She breathes when it breathes and dreams till the milk dries.
Her body is gone. She is only two hands.

Her face to its flanks and lulled by the milk-beats,
She feels her mind empty itself and grow still.
As though the long-borne at last is delivered,
A lightness has lifted her thoughts and her will.

She has the same ease as a sky after thunder,
Sure only of silence and safe, dousing rain,
Or a country that watches its enemies leaving,
And rests before manning its own forts again.

There is something unknown between her in her stillness
And the cow standing there in its soft clover steam,
One chewing the calm of inherited daisies,
And one, even-breathed, on the cud of a dream.

Only the nuclear family with its responsibilities apportioned between men and women could have developed the small farm and it was women who held the whole operation together. They received few accolades outside their families for their courage, stamina and enterprise. When their families grew up and they managed some spare time they banded together in the women's section of the Farmers' Union or (later) the Women's Division of Federated Farmers and in the Women's Institute. Pictured is the first federation council of the Women's Institute gathered at 'Omatua', Rissington, Hawke's Bay, in 1925. The Institute not only supported women's interests but encouraged crafts and arts, holding one-act play performances, for example.

This paean was one of the too few gestures towards the work of women in New Zealand's pioneering society. Only the nuclear family with its responsibilities apportioned between men and women, the way they were then, was capable of developing the small farm, and it was the women who held that unit together with stoicism, common sense and extraordinary stamina. Many, if not most, of the dozens of local historians have been women and have become aware of this oversight. They have noted that most people recorded and talked about the work and public lives of their fathers and grandfathers and had to be prompted to talk about their mothers and grandmothers. Pioneer women produced children with what one historian called 'startling regularity', and many died from childbearing or just plain exhaustion. They received few accolades outside their families for their courage and enterprise.

For many impecunious men and women who chose the wrong land at the wrong time and were inadequately prepared by previous knowledge and experience, the discomfort and work were futile and they never made it through to solvency. Frank S. Anthony, humorist (*Me and Gus*) and novelist (*Follow the Call*), bought land with government assistance after serving in World War One: fifty acres, £45 an acre, £100 down. *Follow the Call* is based on that experience:

> Sometimes I'd get a piece of paper and pencil, and work out how much my cows were going to produce in a season, and how much money I would have saved up in a year. I've kept some of those calculations out of curiosity, and they make me laugh even now. I was only hundreds of pounds out. It was a great experience, sitting in that fire-lit bach, building castles and saying to myself, 'My own boss! No one to give me an order, no one to criticise what I do . . .' My farm was poor and the pasture thin and run out, and as soon as I discovered that the spring was passing me by with just a feeble nod and smile, I switched my hopes on to the summer flush of grass.

Far left: Humorist and novelist Frank S. Anthony made great fun of developing a farm from the Taranaki bush in a series of stories called *Me and Gus,* but in the end the task defeated him.

Left: One of the World War One veterans who won a farm in a ballot, but was defeated by it because the farm was too small and he had insufficient capital, was Rewi Alley, who later became famous for his work establishing and leading a commune in China.

It was a shock to find that my heifers started to go off in the summer, in spite of the summer flush, but after a day or two of worry I took heart again. I had put in three acres of soft turnips for autumn feed, and pinned my faith on the turnips. I relied on them to pull me up and make my herd milk well through the autumn months. By the end of February the grass began to disappear off my section as if melted away by magic, and if it hadn't been for the three acres of turnips my cows would have starved to death. Instead of making them milk it just kept them alive, and every morning down at the factory I used to take the lids off the cans before drawing under the hoist and let some of the beautiful turnipy stink get away . . .

Anthony's hero stayed on for a happy ending. Anthony himself walked off his farm.

The fight against the bush didn't really end until the 1920s, and some historians say the pioneering period wasn't fully over until the 1940s. Late in the nineteenth century and early in the twentieth, a factor in land valuation was proximity to a port and/or a railway service and the state of the access roads. Nevertheless, thousands of people were prepared to try their luck. It wasn't just a search for egalitarianism. Jack wanted to be not only as good as his master but to be his own master, and the small farm in the bush was his best chance. When the big blocks of land were opened up for ballot after the Land for Settlement Act passed by the Liberal Government in 1893, hundreds of men put their names forward and drew fifty acres or more. Most had only a shallow knowledge or experience of farming but they did contribute many skills to a community. They were builders, engineers, horsemen, saddlers, blacksmiths, farriers, bakers, butchers, rabbiters, soldiers, seamen, and a few were schoolteachers and some had experience in running small businesses. Often the one lack was farming experience but they were useful to each other in many peripheral ways, and their common bond was the belief that owning their own land was the route to success and independence. And often it was.

When the Whakatane Cattle Company's station was broken up and lots balloted in 1896 and 1898 – releasing 7000 acres of freehold, followed by

leases covering 12,000 acres of Maori land – more than one person from each family usually put their names down for the ballot to increase their chances. Much of the land was covered by scrub, and drainage difficulties had to be attended to, but it was mostly rich loam soil with very few stumps from forest trees to be blasted out. For most of that decade the settlers tried growing oats, wheat, barley and even sorghum and sugar beet before settling on maize as a main cash crop. Then in 1898 and again in 1900 severe frosts around Christmas time devastated the fields. Local historian Alison B. Heath wrote: 'Every settler with a decent crop of maize was ruined. Blackened stalks filled every paddock and there was nothing left . . . Those with cows still had an income, even if it was a daily chore to milk them.' Thus a dairy company was formed and cows became the main source of income.

Part of the vision of the Wakefields had been village settlement by yeomen – a vision tinted by nostalgia after so many of the yeomen of England had passed into grim city factories in search of wages. Many New Zealand settlers remembered the time in England when a man farmed in a small way if he chose to – that was before the enclosures, and before the more ruthless dispersals by the landlords of Scotland. Some of them carried a loathing of landlordism in their bellies. Until the middle of the eighteenth century much of the land in England was still farmed in open paddocks by cottagers who shared rights to common land, by yeomen who were freeholders, by leaseholders, by 'copyholders', all of them free from eviction as long as they paid their dues. This was part of the intrinsic structure of English society. But it broke down from the mid-eighteenth to the late nineteenth century.

Ownership of land in Britain carried with it political power, which the gentry proceeded to use to wrest the often undocumented legal rights to land from those they considered their lessers. About 4200 Enclosure Acts were passed through Parliament between 1760 and 1845, resulting in ruthless mass evictions from long-held land. The victims became acquiescent farm labourers, or trudged into the cities, or emigrated – some of them to New Zealand. This dispossession seriously deprived ordinary people who also felt deeply that land ownership conferred status as well as security. Writer, historian and philosopher Thomas Carlyle wrote in 1843: 'The land is mother of us all; nourishes, shelters, gladdens, lovingly enriches us all.' From the days of Ancient Greece, the ownership of land gave aristocracy its status.

Many blamed the ills of Victorian society in Britain on the massing together of people in cities, a consequence of the Industrial Revolution. The countryside was considered the ideal, the healthy, the natural habitat of man. The new city-dwellers could trace their troubles back to the day they left their villages and farms. And it was true that large, densely settled cities, a new phenomenon in the world, were dangerous places. The natural increase in population was higher than it had ever been before. Investment, planning and technology could not keep up with the consequences of rapid urbanisation, and its subsequent problems such as food preservation and distribution, clean water reticulation, sewage disposal and the preservation of order. The material and spiritual woes of the world were blamed on the abandonment of the 'natural' life rather than on the disintegration of a villager's tribal, moral and cultural mores.

The belief that the remedy for social and mental ills was to return to the bucolic life was a driving force behind the migration that settled New Zealand. It manifested itself in many ways on arrival, including the quarter-acre section, the siting of hospitals on hills on the edge of town and the building of mental asylums in the countryside. Until well after World War Two, it was a basic tenet

The farmers' Governor-General

Charles Bathurst, Lord Bledisloe, was regarded as the farmers' Governor-General (1930–35) because of his deep interest in and knowledge of pastoral farming. He was a regular speaker at farming conferences and conventions during his term of office and his knowledge and experience meant he was keenly listened to. The most famous trophy he left to mark his stay was the Bledisloe Cup for rugby competition between New Zealand and Australia, but he also donated awards for gardens featuring native plants and for the best Maori farmers.

Bledisloe was educated at Eton and Oxford, qualified as a lawyer but then studied for three years at the Royal Agricultural College, Cirencester. He was elected to the House of Commons in 1910, was parliamentary secretary to the Ministry of Food towards the end of World War One, and then parliamentary secretary to the Ministry of Agriculture and Fisheries, 1924–28. He became chairman of the Imperial Grasslands Association in 1928, and from then on his interest in farming was paramount. He quickly attuned to the New Zealand community during the hard times of the Depression, arranging for his salary to be reduced by thirty per cent in line with cuts to the salaries of public servants, and taking a focused interest in gardening and farming.

His most extraordinary and generous gesture to New Zealand culture was made in 1932 when he bought the property on which the Treaty of Waitangi was signed and presented it to the nation with the national centennial celebrations eight years ahead. For more than thirty years, the Early Settlers and Historical Association of Wellington held an annual Bledisloe Day.

Governor-General Lord Bledisloe took a great interest in New Zealand agriculture and was an advocate of the health-giving qualities of the rural life. He is pictured here getting a garden-tools salute by a strange guard of honour as he arrived to deliver a speech.

On his return to England, he became famous for his Red Poll cattle, kept a dairy herd, farmed pigs and grew grain and other crops. He was elected president of the Royal Agricultural Society in 1946 and the following year visited New Zealand in that capacity on a goodwill mission.

---

of public health policy that gardening, and working in the open air was physically, mentally and emotionally healthier than living in paved urban streets. The Governor-General, Lord Bledisloe, gave the Cawthron Lecture at Nelson in 1932 on 'A Conspectus of Recent Agricultural Research'. He opened his address with the assertion: 'Science is sometimes held accountable for many, if not most, of the world's present ills, such as ultra-mechanisation with consequent unemployment, chemical warfare and irreligion. So it is.' Although he went on to urge 'balance' in attitudes to science and progress, his printed version had a curious footnote that read, 'By a strange irony Science has also contributed to the marked increase of insanity in proportion to population, by intensifying the pace (and consequent nerve strain) of life by its inventions, and by the humane preservation (and incidentally the reproduction) of mentally defective human beings.'

In the earliest years, many settlers found that the idyll of migrant dreams required more land development capital than they possessed, and that the economics of numerous small farms dependent on a domestic market in a new community did not make sense. Refrigeration, though, enabled dairy and fat-lamb farmers who were not too burdened with debt to wrest a living from as little as fifty good acres of land. The operative word, though, was 'good'. Land policy was the crucible of New Zealand politics right through the settlement period, until World War Two. The Minister of Lands in the 1890s Liberal Government, John McKenzie, was the son of a Scottish tenant farmer, who was raised at the time of the crofter evictions from the land. He emigrated to Otago. He was a conservative in all other matters except land tenure. He hated landlordism and believed farms should be owned by those who farmed them, although he favoured leasehold over freehold. He was instrumental in breaking up most of the large sheep runs and placing families on smallholdings. A powerful leasehold movement remained in New Zealand for many years until it was eventually defeated, as it was bound to be in a country that was moulded in the Anglo-Saxon tradition.

Once the smallholders were on the land they nurtured the ideal that theirs was the natural and therefore virtuous way of life. 'Rural life was regarded as essentially wholesome,' says Keith Sinclair in his *History of New Zealand*. 'The morality of townsmen was (and is) suspect . . . To farmers the future seemed to lie between socialistic, loafing townies and hard-working, individualistic, freeholding countrymen.' This credo and its obverse – townies' belief that farmers were all wealthy and politically favoured property-owners – created social and industrial tensions and erupted in violence when it was politically exploited (as in the conflicts of 1913 and 1951).

Farmers did indeed wield enormous political power and were even permitted a form of gerrymandering for many years in the form of a 'Country Quota'. At one time, if the President of the Farmers' Union or, more recently, Federated Farmers, wanted to see the Prime Minister, the Prime Minister's secretary would possibly say, 'What time suits you?' That situation lapsed in the early 1970s, and was dramatically underscored by the abandonment of awareness of farming's importance when the Labour Government of 1984 pulled their subsidies away. The power had passed on to more disparate, large urban groups. Interestingly, since then farmers have simply got on with their jobs and proved that the country is still dependent economically on their enterprise and skill.

Are New Zealanders still emotionally attached to the land then? They are ambivalent. While thousands are migrating from the suburbs into downtown apartments, many are buying beach houses, lifestyle blocks and boats to get away from the urban rush.

The Minister of Lands in the 1890s Liberal Government, John McKenzie, hated landlordism because he was raised at the time of the crofter evictions in Scotland. He believed farms should be owned by those who worked them and was responsible for breaking up most of the large sheep runs and placing families on smallholdings.

CHAPTER TWELVE

THE MARRIAGE OF FARMING AND SCIENCE

UNTIL THE END of World War One, farming in New Zealand was controlled by practical men who had little time for the intellectual pursuits associated with scientific research. They observed problems and did their best to solve them in a commonsensical way. Commercial farming on any scale began with running sheep over native tussock in the 1840s and 1850s. As L. G. D. Acland put it in *The Early Canterbury Runs*: 'The runholder's practice was simple at first. He put his sheep on the run and tried to keep them there. He marked the lambs when necessary and shore the sheep once a year . . . He tried to breed up his flock as fast as he could. Until about 1868 when all the runs were fully stocked there was always a good market for store sheep.'

Many of the squatters were intelligent, well educated, and possessed at least some capital. Acland said that among them were 'baronets, younger sons of good families, soldiers, sailors, parsons, lawyers, tradesmen, shepherds, farmers and . . . several foreigners'. But no matter how intelligent and experienced the runholders were, their lives were relatively straightforward contests against snowstorms, or to contain diseases like scab. Successful all-round farmers like John Grigg used their experience and repositories of knowledge gained in England to breed up their animals and grow their crops and they did use records to refine their work. The farmers carving farms from the bush in the north acquired their knowledge from shared experiences. But what arable and bush and sheep farmers did recognise and accept readily were advances in mechanisation that reduced costs and gave relief from the back-breaking labour on the land.

Otherwise, interest in science, or even the toleration of it, was hard to generate among the mass of practical, grafting farmers who believed all problems bowed before hard work and common sense. From that same outlook arose the indifference towards agricultural education at secondary school or tertiary level unless it could be seen to be wholly utilitarian. Towards the end of the nineteenth century, a Hawke's Bay farmer and Member of the Legislative Council, the Hon. Mr Wilson, in condemning the Lincoln School of Agriculture, said he did not agree with 'making farmers through a course in chemistry'.

But in the early years of the new century it became apparent that New Zealand soils were not as naturally fertile as had been assumed and that soil deterioration was resulting in erosion and waning productivity. Between 1840 and 1910, about four million hectares of forest, fern and scrubland was grassed by surface sowing after burning the natural vegetation. Immediate results from the potash deposits were good but the assumption that the pasture would hold on and flourish became observably untrue in many cases and science was the wizard to which afflicted farmers looked.

In 1905 a Mr P. Pattullo, of Otago, said, 'The land does not now smile with harvest when tickled with a hoe as it did twenty years ago . . . We find also that our pastures do not hold in grass so well

A decade after a government leader had condemned tertiary courses as useless for producing good farmers, this 1910 class at Stratford District High School examines experimental pasture plots. Few, if any, of these boys would have moved on to Lincoln College, the only agricultural university. Interest in science was hard to generate among early farmers who believed all problems bowed before hard work and common sense.

as they used to . . . Land is getting dearer, expenses heavier, and it will require more attention, more knowledge, and more trouble on the farmer's part to grow crops successfully than in the early days.'

In response came the first assault on agricultural problems by science, with work mainly by chemists and botanists on soil analysis, the variable requirements for fertilisers, and the productivity and longevity of available types of pasture. Rock guano was brought in during the 1860s and 1870s but the need for regular fertiliser application was underlined by the director of Canterbury Agricultural College at Lincoln, W. E. Ivey, who imported a shipment of superphosphate in 1880. The following year commercial production of the fertiliser began in Dunedin.

B. C. Aston, a Department of Agriculture chemist, and others, made profiles of New Zealand soils and found great variation in textural properties and in the content of organic matter. Aston was also involved, along with Edward Vaile, in solving, more or less by accident, the animal health problem of 'bush sickness' that stalled the development of large areas of the North Island's central volcanic plateau. Vaile owned 53,000 acres of the pumice country and managed to keep his stock in good health by using limonite, an iron compound, on Aston's advice. Aston believed the deficiency was iron, but by the 1930s he and Vaile realised that the actual problem was lack of cobalt, of which there was enough in limonite to make a difference. The sickness was completely countered by the application of cobalt. Trace elements, such as

Dr Leonard Cockayne, shown here in 1897 (centre), was the first of a number of research scientists who did work that was internationally respected in the area of breeding pasture types specially adapted to New Zealand conditions.

molybdenum and copper, were found to be missing in other regional soils.

By the 1920s the Wheat Research Institute at Lincoln College had intensified interest in plant breeding and selection. An early inspiration was Dr F. W. Hilgendorf, botanist and zoologist, a versatile teacher and researcher whose work with cereals brought gifts to New Zealand farming. Dr Leonard Cockayne, and later Sir Bruce Levy, steadily edged forward with internationally advanced work on the selection and breeding of pasture types and the use of nitrogen-fixing clovers as a valuable, cheap alternative to nitrogenous fertilisers. Levy's mana among farmers ensured that his good advice was widely spread and earnestly taken. It was slow, undermanned and underfinanced work but, in the end, wonderfully effective. By 1932 Lord Bledisloe could say of grasslands work: 'There is no field of research and experiment more closely affecting New Zealand husbandry and none in which New Zealand scientists and farmers have more skilfully pointed the way to the rest of the Empire.'

What was lacking, though, was an overall plan for the support of industry by science. The New Zealand Institute, a scientific organisation founded in 1868, made some suggestions about the organisation of scientific research. Then, in 1925, the government asked the director of England's Department of Scientific and Industrial Research (DSIR), Sir Frank Heath, to visit New Zealand and advise on how best to organise science to support industry. The government responded to his report with the Scientific and Industrial Research Act of 1926, which set up the New Zealand DSIR and signalled the beginning of governmental patronage of scientific research that lasted for more than fifty years.

Heath pushed hardest for science to assist the primary industries, starting with dairying as the most urgently in need. Dairy exports the year he

Pasture care is more important today than ever. Murray Stratford and son Nick reseeding pasture on their farm in Progress Valley in the Catlins.

visited were worth £16 million and no one could escape the conclusion that in the foreseeable future the economy was going to stand or fall by the export of wool, meat and dairy products, and also some crops and pip fruit. Heath suggested an institute of dairying be set up and used as a model for other organisations to promote the production and processing of grain and grass crops, fruit, sheep and cattle, and meat and wool.

The leaders of the dairy co-operative movement were already aware that in Denmark and the United States laboratories were exploring the potential for milk products. In 1921 the biggest amalgamation of co-operatives in the country had come together as the New Zealand Co-operative Dairy Company and established a laboratory in Hamilton for routine testing directed towards turning out a wide range of standardised, high-quality products. The company was already manufacturing butter, cheese, casein, dried milk and dried skim milk. Companies in Taranaki had formed a professional association, the Federation of Taranaki Co-operative Dairy Factories, and this group set up a laboratory in Hawera in 1925 with help from a Department of Agriculture grant of £1000 a year.

The government decided to establish Massey Agricultural College at Palmerston North and directed it to concentrate primarily on dairy issues. Nearby was the Dairy Research Institute which soon affiliated the laboratories in Hamilton and Hawera. An experimental dairy factory was set up. However, much of the energy and enthusiasm waned when the economic depression stalled capital investment in all industrial activity.

The next big move came with the appointment in 1943 of Dr C. P. McMeekan as director of the Animal Research Station at Ruakura. McMeekan was a foundation student at Massey who later graduated with a doctorate from Cambridge University. He was an extrovert who listened to the problems of the ordinary farmer and spoke his language in reply. His communication and promotional skills were so strong that he became a celebrity scientist and administrator. McMeekan transformed what was essentially just a government farm into a property on which empirical science could be applied to the basic problems of animal breeding and production. In the process, he assembled a cast of scientists, many of whom became known in every pastoral farming country in the world – John James working on artificial insemination, Lindsay Wallace on animal nutrition, Wattie Whittlestone and Doug Phillips on the whole process of milking cows, John Hancock and Alan Carter on genetics. The stellar decade of work at Ruakura was 1948–58 when it was acknowledged by some scientists in other countries as the best animal-research establishment in the world.

At the same time, the Waikato produced a number of extraordinary farmers. One was Jack Ranstead, who assembled an incomparable private library, and from that and his work on his farm developed both theoretical and empirical knowledge of animal breeding and husbandry that ranked him among the top world experts. His father, William Ranstead, had been a friend of the famous English writer and artisan William Morris, and after a visit here in 1899 returned to England determined to test his Fabian faith by emigrating with a group of like-minded, high-minded people to establish a

Utopian communes were all the rage in the nineteenth and early twentieth centuries. William Ranstead (seated in front of the man with his hands in his pocket) with a party of the skilled craftsmen soon to emigrate with him, was a Fabian and friend of English writer and artist William Morris. Four years later, in 1896, he brought a party of 200 high-minded settlers to New Zealand on four ships in 1900 and the number swelled to 2000. But it was Ranstead's son, Jack, who made perhaps the biggest contribution to this country. He became a leading advocate and active participant in animal production research.

Utopian settlement in New Zealand. Utopias were all the rage at the time. The party of 200 came out on four ships in 1900, and later arrivals swelled the group to two thousand. This Utopia, like almost all of them, collapsed almost as soon as it began, so Ranstead eventually settled at Matangi, near Hamilton. Jack Ranstead, one of Williams' five sons, gained a diploma in dairy farming at Lincoln College and began his intellectual and practical interest in animal production. He became the first president of the New Zealand Animal Production Society during the early years of World War Two and was an active participant and patron of research. A neighbour of the Ransteads at Matangi was Dr H. E. Annett, who settled in this country in 1927 after twenty years in the Indian Agricultural Service. For many years Annett conducted fertiliser and rotational-grazing experiments on his property at his own expense.

Before the flowering of Ruakura, practical, applied animal research had been submerged by the breed societies which kept a reasonable standard of performance with the most widely used breeds, but relied too much on the often irrelevant qualities of conformation and trueness to breed type. British judges were frequently brought in to agricultural and pastoral society shows and judged long-established New Zealand adaptations of cattle and sheep breeds according to the conformation standards of British types. With scientific backing, the mystique of conformation was almost totally supplanted by the pragmatism of production recording.

Of course, some clever practical farmers had known this truth for many years. A leader of the National Dairying Association from 1899 to 1921, Mr T. G. Harkness, told the association conference the year he retired:

Do you know what I think is the best work I ever did for the industry? When I stood on this platform and read to you a paper on how to improve a dairy herd long before anyone else had taken the matter up. I said, 'Don't waste your strength on feeding a cow that is not profitable, but see, every one of you, that you have a herd that can do 250 to 300 lb of fat and then you will be able to meet the difficulties that are facing us in the near future.'

That year, 1921, the average fat per cow was 174 pounds.

And yet by the 1950s productive performance was still a secondary issue with some. McMeekan and his scientific staff deflated breed society pretensions with Ruakura's unique (at that time) commercial herd-size experiments on animal breeding, production and feed conversion, grazing techniques, the milking process and other parallel issues that revolutionised dairying by 1960. By the 1970s a Ministry of Agriculture pamphlet on commercial sheep flock improvement said simply: 'After an eye appraisal for wool or physical fault, the farmer should choose a ram on fertility rating, fleece weight and bodyweight recordings.'

The Ruakura group made the paddock their laboratory. If research projects did not apply directly to the business of livestock farming under conditions prevailing in this country, they did not undertake them. They aimed at achieving the highest production per acre at the lowest cost to the farmer.

Scientists at Ruakura undertook a wide range of other work involving farm machinery development, and animal reproductive physiology, nutrition and

Dr Percy McMeekan turned Ruakura Animal Research Station into one of the world's outstanding research institutions and became a celebrity scientist and administrator.

health. One of its most publicised achievements, which was a long time coming, was the discovery of the cause of a disease known as facial eczema. It was caused by a toxic fungus that grows on pasture following rain and high humidity in the late summer and autumn. First noted in 1882, it became a nagging problem for farmers, especially in the high-production regions where warm rain in autumn and summer kept the grass growing. Finding the cause became a matter of serious national concern in 1938 when millions of sheep and cattle were hit by the fungus which attacked the bile ducts and liver, and caused the visible symptoms of a sunburn effect and scabs on the animal's face. Scientists worked on the problem assiduously as pressure mounted to find the cause. But it eluded them for twenty years, mainly because they had dismissed too early the likelihood that the disease was fungal in origin. One day, a scientist, Campbell Percival, noticed a black dust on the blades of a mower, collected some, identified it as the spores of a fungus, cultured it and fed it to guinea pigs. The search was over and a preventive was on the way.

The eminence of men like Levy and McMeekan effected a strong change in attitudes towards agricultural education after World War Two. At secondary level, the agriculture courses had long been regarded as a refuge for dullards. At tertiary level, educationists seemed to lack the will to move and farmers the drive to push them. Although the Lincoln School of Agriculture was the first in the southern hemisphere when it opened in 1880, it soon lapsed into an educational coma, and a report in 1925 opined, 'The present position makes our task (of planning the future of agricultural education) especially difficult; Canterbury, a Special School of Agriculture, without a professor; Victoria College, with a professor and ten students but no School; Auckland, with a professor and neither students nor School.'

After McMeekan left Ruakura in 1962, many Lincoln and Massey graduates became station scientists and continued to make great progress on animal production at Ruakura and a number of other government research institutions around the country. Their efforts helped keep New Zealand up among the top agricultural countries in the world.

In 1996 in Scotland a team of scientists, led by Professor Ian Wilmut, successfully cloned the first mammal, Dolly the sheep, from an adult cell. Controversy immediately followed as scientists and lay observers discussed the ethics of creating new animals from the cells of others, with possible unforeseeable consequences. Always in the background of the discussion loomed the fear of unscrupulous adaptation of cloning for creating human beings. After Dolly, other mammals were cloned, including horses and bulls. While animal cloning is still a difficult and costly procedure, AgResearch scientists believe it will eventually have a place in pastoral farming in New Zealand. By 2006 thirty AgResearch staff at Ruakura had been investigating animal cloning for some time.

The Marriage of Farming and Science 147

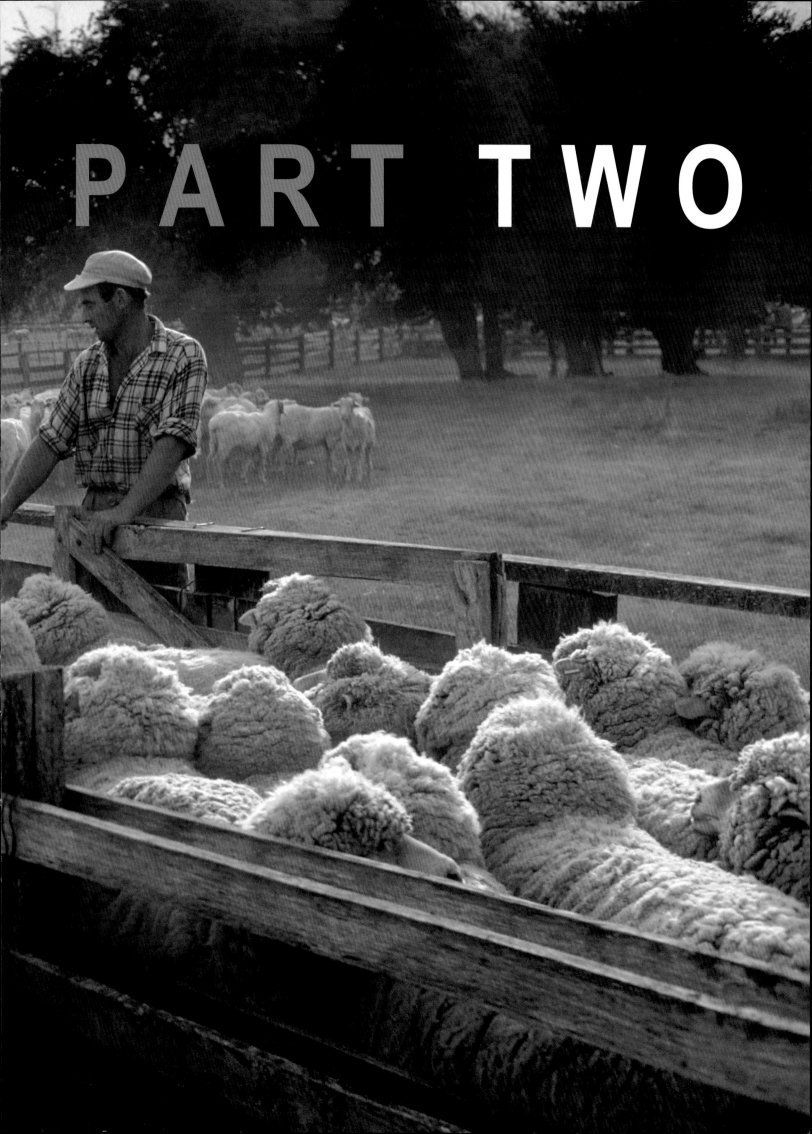

PART TWO

CHAPTER ONE

THE FARMER'S WAR, AND THE AFTERMATH

Between the wars, New Zealand ran marketing campaigns in Britain, especially during the Depression when Britain began moves to place tariffs and quotas on meat and dairy imports.

The Roslyn Worsted and Woollen Mills in Dunedin produced fabrics from New Zealand wool for generations, including for uniforms and blankets for troops during World War Two, using more women workers than ever before.

NEW ZEALAND paid dearly in blood during World War Two with an estimated 75,000 people joining the fight against Germany in the Middle East and Europe, and against Japan in the Central and North Pacific. About 36,000 died, were wounded or imprisoned during the nearly six years' struggle. In July 1942 mobilisation peaked at 154,549, including 8500 in women's services but excluding the Home Guard and Auxiliary Patrol Services. It was a colossal effort on the part of a small country. But, while many cities in Europe and Asia were razed by aerial bombing and artillery fire, New Zealand's infrastructure was not damaged – even though it was neglected in most of the years through the stuttering recessions of the 1920s, the big Depression of the 1930s and the war.

Vigorous farming continued during the war years and production remained fairly steady despite shortages of labour and fertiliser, and lack of investment. Producing wool for uniforms and blankets, and traditional meat and dairy products to feed embattled Britain, our own servicemen overseas and American troops in the Pacific was seen as part

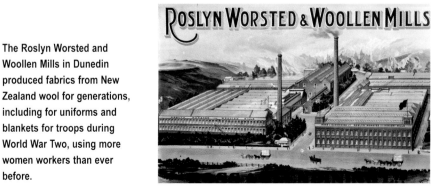

of the war effort. A scheme called 'Manpower' absolved men from conscription to the armed services if they were working in essential industries such as farming, although many farmers enlisted and left family members, including fathers, mothers and wives, to work the land.

Manpower was also used to direct workers not liable for conscription to industries deemed essential. Land Girls were recruited to help keep farm production up. Although resistance to an organised force of Land Girls was expressed at first by some politicians and farmers, nearly 3000 members of the Women's Land Service (WLS) worked on farms after the service was formed by government regulation at the end of 1941, a few months before full Manpower controls were put in place. Although service was voluntary at first, members of the WLS could be compulsorily assigned to farms by Manpower officers from the beginning of 1944. Just over half the Land Girls worked on non-family farms.

The Labour Government had formed a marketing department in 1937 which controlled the export of dairy produce. This department and other centralised marketing controls already in place enabled the government to mobilise primary products as soon as the war started and export them to Britain at set prices. These guaranteed prices relieved farmers of concerns about revenue even though the meat and dairy prices were generally below world levels until a few years after the war. So the contribution of New Zealand farmers to the war was considerable.

Young women as well as young men joined the Maori exodus from country to city during and after World War Two. Women gather in the lounge of the Pendennis Maori Girls' Hostel in Wellington in 1954, working at handicrafts after a day at their city jobs.

During the war, the Maori exodus from the country to the city began. It gained momentum in the immediate post-war era. For decades many had lived at subsistence level in rural areas. Despite land assistance programmes nurtured by Sir Apirana Ngata in the 1920s and 1930s, the limited land base and lack of development capital following the confiscations of the nineteenth century and continued purchases through the Maori Land Court meant they were never able to regenerate the drive and productivity of the years before the wars of the 1860s and 1870s. The Maori birth rate at the time was three times higher than for Pakeha. By the end of the war, 17,000 Maori lived in urban areas – more than twice as many as ten years previously – and during the twenty years to 1971 urban Maori numbers grew from 20,000 to more than 130,000. Many of them worked in processing plants for pastoral industries, especially as butchers in freezing works. As automated plants took over processing and replaced the big export works, thousands of Maori lost their jobs.

A wartime bulk-purchasing agreement between Britain and New Zealand for primary produce continued until 1954, so the incentive to seek out new markets remained low. Cheap synthetic fibres developed during the war came onto the market and raised some concerns about the long-term viability of wool but they disappeared, temporarily, when the Korean War hefted demand to such a degree that the New Zealand Government temporarily froze some of sheep farmers' incomes to prevent inflation running amok.

They were good times, the 1950s and 1960s, for traditional primary commodities and thus for the economy. Revenue was bolstered for the first time by forest products, including pulp and paper from the mill at Kawerau, in the Bay of Plenty, which came on stream in the 1950s. New Zealand was then one of the richest countries in the world per capita. In 1969 the National Government appointed a Royal Commission into the future of social security and the welfare state with the intention of extending it in a rational way. The family benefit was doubled. The Royal Commission's report was disappointingly flatulent but it contained valuable statistics. Accident compensation, a no-fault insurance scheme for workers, was introduced and then extended over the following few years to cover all citizens.

In the immediate aftermath of the war, most commentators considered the feeding of a hungry world with a resurgent population augured riches for the future of food-producing countries. Farm productivity had increased at varying rates since the early 1920s. The number of people working on the land gradually diminished here as well as in other developed countries, while production volumes climbed as a result of increasing professional skills and technological advances. About one-eighth of the New Zealand labour force was employed in agricultural production by the 1970s, down from about a third at the beginning of the war.

At the end of the 1960s most New Zealanders could see no end to the golden years, but by then the future was blurred for careful observers by the slow growth of the British economy on which prosperity depended, by increasing protection for British farmers, by the near certainty that Britain would join the European Common Market, and by the abysmal lack of marketing information available in New Zealand with which to diversify products and markets. The country's economic outlook changed abruptly before the 1970s were very far advanced. Even before Britain finally clinched the 1973 deal with the EEC, international oil prices surged and knocked developed economies from their relative equilibrium. Wool prices slumped again, this time beginning a decline from which they have never really recovered. The terms of trade fell by a staggering forty-two per cent by the time of the 1975 election and social dissension erupted as unemployment escalated and damaged the national confidence.

The belief gained momentum and finally prevailed that growth would come only through manufacturing and financial and tourism services. Although New Zealand had carefully negotiated quotas with Britain for meat and dairy products for a fairly long period, other developed countries – and economies like Japan's, growing phoenix-like from the ashes of war – were unambiguously refusing unencumbered entry to agricultural products. The Norman Kirk/Bill Rowling Labour administration borrowed and hoped. The Rob Muldoon National Government from 1975 failed to accept the economic realities of world changes and floundered for a decade, mostly trying political nostrums. Having condemned borrowing by Labour, Muldoon borrowed even more – and waited Micawber-like for something to turn up.

Some progress was made. A few financial controls were eased and the New Zealand-Australia Free Trade Agreement (NAFTA) stimulated the export of manufactured goods to Australia. NAFTA evolved into the free trade of Closer Economic Relations (CER). The diversification of both products and markets began, but slowly.

The economy was in deepening trouble by the time the 1980s arrived. As conditions deteriorated, wage and price controls were adopted, vainly, to damp down inflation. The public was restless as Prime Minister and Minister of Finance Muldoon tried in his cajoling, often bullying, way to conduct the economy like an orchestra, playing his own arrangements in a bid to get the country back into tune. To no avail. It was all temporising, and when change arrived precipitately in the mid-1980s farmers were hit first and hardest.

Perhaps livestock incentive schemes and supplementary minimum prices (SMPs) for meat producers best illustrate how far the business of farming was removed by price supports from the chastening influence of prevailing international prices. In 1978 when inflation was almost out of control and the New Zealand dollar was artificially high against other currencies, experiments with various price supports ended with the government imposing SMPs to insulate farmers against rapidly growing costs. The plan was to give farmers confidence in a future in which the commodities price cycle would once again swing up and balance out losses – as it always had done, sooner or later, in the past.

Opposite: The contribution to World War Two of New Zealand farmers was considerable. Not only did they take lower commodity prices but many served overseas in the armed forces. Food and wool production at home was, in many cases, left in the hands of men who were ineligible for conscription, and women who became what were commonly called Land Girls. At first, the government was reluctant to even enlist women but by 1943 many were hard at work in the country. Pictured (top left) are Carol Sladden and June Matthews in the dress uniform of the Women's Land Service; a dairy worker in the Rotorua district (top right); and in Hawke's Bay, a woman helping to ear-tag a lamb (bottom). All pictures were taken in 1943.

Prime Minister Robert Muldoon and the Australian Deputy Prime Minister, Doug Anthony, begin talks in Wellington in 1982 on CER (Closer Economic Relations), which had its genesis in the New Zealand-Australia Free Trade Agreement of the previous decade.

Although the trigger point for SMPs did not immediately arrive, the system became a bulwark against despair. But the trouble was that stock numbers climbed artificially as supply was disconnected from demand. Facilities for handling the kill and distribution systems expanded and new markets could not be found fast enough – not at satisfactory prices anyway.

Price supports allowed a traditional attitude to become a delusion. Bulk exporting, with prices often agreed between British and New Zealand governments, had been the standard for so long the primary industries and the government found it hard to get a grip on the concept of producing food for specific customers. As late as 1983 the Meat Producers' Board commissioned a business consultancy, Glendinning and Associates, to report on the distribution and marketing of lamb in Britain. The result was damning. New Zealand lamb was 'the worst presented and least attractive commodity available in the supermarket'. Without slick, modern packaging and presentation the chance of getting top returns for meat from discerning consumers was negligible.

Meanwhile, the national flock eventually exceeded seventy million. SMPs for meat were more than twenty per cent over prevailing prices and by the end of the 1983–84 season had cost $595 million – about $450 million of that to top up lamb prices. Marketers bustled around the world looking for new deals to take produce gradually being shut out of Britain plus the new meat mountain created by subsidies. A lump-sum support that replaced SMPs in the following season cost a further $130 million. One estimate is that in their biggest year, 1982–83, the whole range of farm subsidies cost well over $1 billion.

Worldwide, those who work the land tend to have a more conservative cast of mind than urban people and, politically, in a sense relative to the

Trade panic during the Depression

Nearly forty years before Britain joined the European Common Market, the Depression of the 1930s caused the first major disruption of trade with New Zealand. The British attempted over five years to impose quotas, tariffs or levies to protect their own primary industry. After ninety years of a laissez-faire approach to agricultural trade, Britain negotiated new five-year agreements at an empire conference in Ottawa in 1932. At the time, Britain was taking all of New Zealand's cheese, ninety-nine per cent of its butter, lamb and mutton and ninety-eight per cent of its beef. Some primary products that had established growing markets in the United States and Canada had suddenly been shut out by prohibitive tariffs a few years before. This almost total dependence on the British market engendered a measure of panic among New Zealand legislators, already trying to cope with internationally depressed prices. One New Zealand commentator said it seemed likely that the Ottawa conference, 'or perhaps we should say the Depression, will be long regarded as a turning-point in imperial relationships'.

The five-year agreement provided for the United Kingdom to impose a duty on dairy products, albeit with preferential margins for the Dominions; to bring products under a general scheme of quantitative control; and for New Zealand and Australian mutton and lamb shipments to be kept static – effectively a quota. The Dominions on their part agreed to maintain preferential margins in favour of British products.

In 1933 the British signed three-year trade agreements with Denmark, Sweden and Argentina which gave British goods access to these countries in return for entry of certain farm produce. Because of the impact on empire preferences, the agreements were called the 'Three Black Pacts' in New Zealand. A year later, the British Government offered domestic subsidies for domestic milk used in cheese production and, in 1935, it announced that its long-term meat policy was to protect its domestic industry with 'a drastic reduction of imports of meat . . . from all sources'. A levy would be imposed on all meat imports, with preferences for the Dominions, the money to be used to help British producers.

The New Zealand Government was so alarmed by this development that it replied that such a levy would severely penalise its mutton and lamb producers and would involve the country in serious economic loss. The reply noted that the economic development of New Zealand depended on export returns for primary products which, in turn, affected its ability to service the charges on British capital investments – a broad hint to bankers and other investors. The government's retort that the idea of a levy was especially repugnant to meat producers and the people of the Dominion was said to be the first time a New Zealand government had taken an aggressive public stand against the 'Mother Country'.

As the world eased its way out of the Depression, the levies were never introduced, and by 1936 reasonably satisfactory arrangements were made between Britain and the Dominions. The war and its aftermath meant Britain's door was once more opened wide for New Zealand food until it joined Europe.

national political centre. New Zealand farmers were no different. They could understand and accept mechanisms of production and price that had been implemented at various times over the years to level out returns between the peaks and troughs of supply and demand; but as governments and bureaucrats, including their own, dithered in the face of impending changes to New Zealand's international trade in the 1960s and, particularly, the 1970s, many farmers became increasingly uneasy at the growing level and type of intervention. They went along with it because subsidies were seductive, but with accumulating discomfort.

What had become abundantly clear to most economists and many farmers was that old devices were being employed to fix new problems. Sheep numbers were so inflated by the 1984–85 season that when the end of the supports was in sight the lamb kill was a record thirty-nine million and the sheep kill nearly eleven million. Two years later, when the rug of SMPs had been yanked away, the lamb kill was down by six million. Marketing teams had fanned out around the world with faint hope that they could sell this mountain of meat. But it would be nearly ten years before the *Economist* could write: 'They may be out-numbered 16 to one by sheep, but New Zealanders do not follow the flock. In economic policy they lead it.'

The Labour Government that gained power at the end of 1984 devalued the currency and declared it would end subsidisation schemes by the end of the 1985–86 season, from which time primary producers would have to confront world markets with a strictly commercial stance. The intention was to deregulate the country's economy as quickly as possible, and because that economy was anchored in agriculture farmers took the first and biggest hit. The new government was ideologically in favour of free markets, and while something had to be done to more sensibly align the country's economy with the reality of world prices, the suddenness was draconian. Cabinet's justification was that, as Macbeth said, 'If it were done when 'tis done, then 'twere well it were done quickly'. The government was aware that fulminating public and political opposition would make the policy hard to implement with such dispatch if it was opened for discussion.

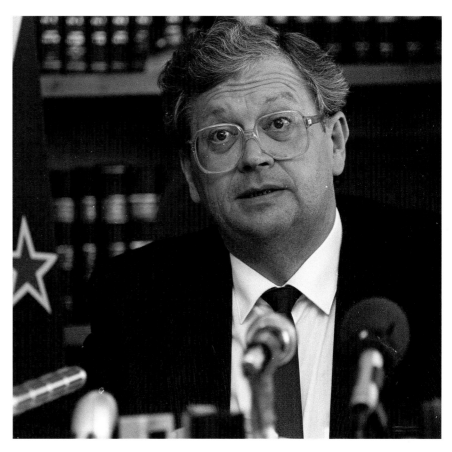

The policy bore down harshly on farmers, especially those who had recently taken on debt in the tradition of young smallholders who spend a lifetime working towards high equity. They paid an exorbitant price for the shilly-shallying of governments over twenty years. The new administration understood the electoral power of rural voters would not be enough to unseat it. Only when urban New Zealanders felt the shock of ineptly introduced deregulation in late 1987 did the government's support drop away.

Furthermore, any services previously provided by the Department of Agriculture that could be construed as financially supporting farmers were stripped away. In 1987 the Department's advisory services were withdrawn and advice was privatised. Inspection fees and adverse-events financial support were removed; and funding was even pulled from government-funded research. All this support was taken away at a time when farmers' incomes were falling after an initial boost brought about by the dramatic currency devaluation. So the 1986 season delivered bad weather, the consequences of a twenty-eight per cent devaluation of the dollar, high interest rates and a future obscured by absolutely no further protection for farmer income.

Prime Minister David Lange announces the dismissal of Roger Douglas from Cabinet in 1988 because he felt economic reform was moving too quickly. The stock market had crashed and urban New Zealanders were becoming disillusioned with the pace of change. But it was too late for farmers who were suffering from the 'cold turkey' withdrawal of all economic supports.

Following pages: Ostriches on Limbo Thompson's farm, Canterbury.

Heavily indebted farmers who had had no hint before 1984 that this would happen were immediately in trouble; and for many, mortgagee sales were the only recourse. One result was a jump in farmer suicides. Thousands marched on Parliament in May 1986, burning effigies of leaders. Topdressing planes buzzed the Beehive with flour bombs, demonstrating that not only farmers but all the ancillary industries were suffering. The effects juddered through many industries and damaged small towns already suffering from the improved roads and cars that could get people quickly and frequently to nearby cities.

Domestic deregulation freed up the long-protected egg and wheat industries, and some competition was brought into town milk supply when a consumer price subsidy was withdrawn along with zoning restrictions on supply and distribution.

In the long run, free-market agriculture was in New Zealand's interests, even though it still bumped into barriers erected against primary products in most other developed countries. But the peremptory nature of the changes seriously damaged some sectors of the community, farmers and urban workers alike. To contradict Macbeth, 'If it were to be done at all, 'twere best to be done well but not too quickly.'

But to see this tumultuous period in its proper social and political context it should be remembered that conventional wisdom considered traditional pastoral farming would not continue to be the foundation of the economy it had been for more than a century. Having to find its true level of value, farming should be allowed gradually to slump to a vastly lesser role if New Zealand was to survive as a First World country – or so it was deemed by most economists and politicians at that time. One Cabinet Minister even described agriculture as a 'sunset industry'. This attitude and the continuing urbanisation of the country, with city voters growing in numerical strength, had ended the power of the farming lobby which had been pre-eminent since just before World War One. The President of Federated Farmers could no longer phone the Prime Minister and expect to be pencilled in for that day or the next. He was now on the B list.

However, many young farmers sensed a kind of freedom in their dissociation from government, a freedom from what had become an accustomed need to lobby. Bulk marketing to a single major destination required some government control and little true initiative from producers beyond negotiating with politicians over the shape of that control. Now they had a need to respond directly to what the people in other countries wanted from them.

The result was the blooming of a long-latent entrepreneurial spirit. Those who dealt with the big traditional products – like butter, cheese and other derivatives of milk, and meat and wool – examined and changed the way they went about their business. Others tried deer farming, the first in the world to do so, goat farming for milk and meat, even ferret farming for fur, ostriches for meat, and other enterprises, all seeking a niche somewhere in either the domestic or a foreign market. Some were successful; some failed. Fruit, berryfruit, flower and vegetable growers who may have been preoccupied by local consumption raised their vision and built up export volumes, using air freight effectively, and slipping into seasonal supply holes mainly in Pacific Rim countries. They looked around for new products.

The glamour stories of diversification were those of kiwifruit, grapes and deer. Early kiwifruit growers made millions and set everyone to fossicking for big ideas, big opportunities. Grapes now pattern thousands of hectares with their long, green grids in many parts of the country, but most famously in Marlborough, Central Otago, and the east coast of the North Island. These grapes in classical varieties feed a wine industry that, over twenty years, has rampantly accompanied the extraordinary culinary revolution. Deer farmers were having a hard time in 2005 but they had added another dimension to pastoral farming.

No doubt was expressed that the New Zealand economy needed a broader base but commentators clutched at straws, insisting that farming could not save New Zealand and that attention had to be focused on information technology, financial services, niche manufacturing and tourism if it were to remain a First World country.

Well, they were right about tourism.

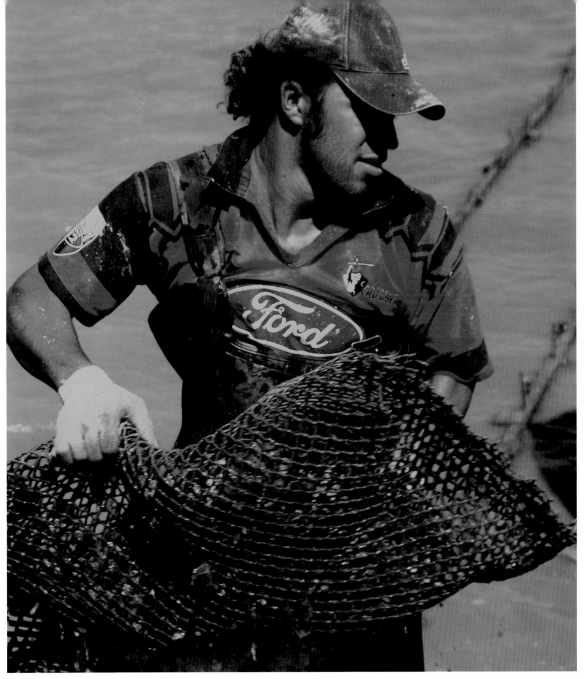

Top left: Aquaculture has become a major industry in recent years in the search for diversity of food production. The two most profitably farmed species are the green-lipped mussel (mostly in Nelson/Marlborough, the Hauraki Gulf and the Far North) and the Pacific oyster (in eastern coastal waters of the north of the North Island). The annual mussel crop is worth $140 million and oyster production totals $45 million, both more than the value of the salmon crop. Stewart Te Tamaki lifts a bag of oysters onto a barge at Clevedon Coast Oysters, near Auckland.

Bottom left: Alpacas look as though they were designed by Dr Zeus but have been domesticated in their native South America for thousands of years. Farmed for their wool, the three thousand, or thereabouts, friendly Alpacas in this country need feed supplements to thrive on pasture. Here is a young competitor with his alpaca at a Hamilton A & P show.

CHAPTER TWO

CAST OFF AFTER NINETY YEARS

In the early 1970s, when Britain's long flirtation with Europe was finally consummated, New Zealanders, including farmers, were gripped by a kind of delayed disenchantment. Suddenly they grasped that Britain – for whom they considered they had sacrificed so much during two world wars – had cast off the mooring lines of its offshore farm after ninety years.

What then gradually became apparent was that the burgeoning post-war economy of Japan would need to become an alternative market for New Zealand. This was anathema to the generation that had not long before demonised the Japanese as an unforgivable, inhumane enemy. Moreover, faced with the need to sell their pastoral products to discerning consumers in diverse markets for the first time, they realised they had absolutely no skill or experience in marketing. Domestically, business had generally been centrally directed, and international trade had been a matter of negotiating deals with British distributors.

Prime Minister Keith Holyoake had accepted Britain's retreat behind the restrictive quota and tariff walls of Europe as inevitable for most of a decade and reacted in a way that was quintessentially his own. With a leadership style that seems retrospectively so attractive in the context of more recent behaviour, he sought consensus by setting up a National Development Conference. Industrial sector committees met and, backed by the efficient and diligent senior public servants of the time, trawled for essential data from their sectors and made recommendations to the conference. The conference was held in 1969 when the country was having its first serious balance of payments problems in more than ten years.

The information presented to the National Development Conference was substantial, relevant and valuable, and many of the recommendations provided a base for action that was not taken for years because of a change of government in 1972 and an economy rocked by oil shocks. For example, the principle of adding value to primary products in this country instead of bulk exporting without further processing was reinvented by many marketers every few years after that 1969 conference. Indeed, 'value-added' remained something of an economic catchword for more than thirty years. The problem then – larger even than it was in 2005 – was that most countries, including Japan, had barriers against importing processed goods, preferring rather to bolster their own economies with the processing stage. For instance, it was easy to export logs but not timber products.

The remedies taken by farmers, by a generation of marketers and, erratically, by governments was impressive and saved New Zealand from economic disaster. Farmers and their governments had known for a long time of two principal dangers perennially facing New Zealand: first was extreme reliance on the British market for produce on which the very existence of the New Zealand economy depended; the second was bulk supply of produce to distributors in London for them to do the consumer marketing on the country's behalf.

From the beginning of commodity exporting in

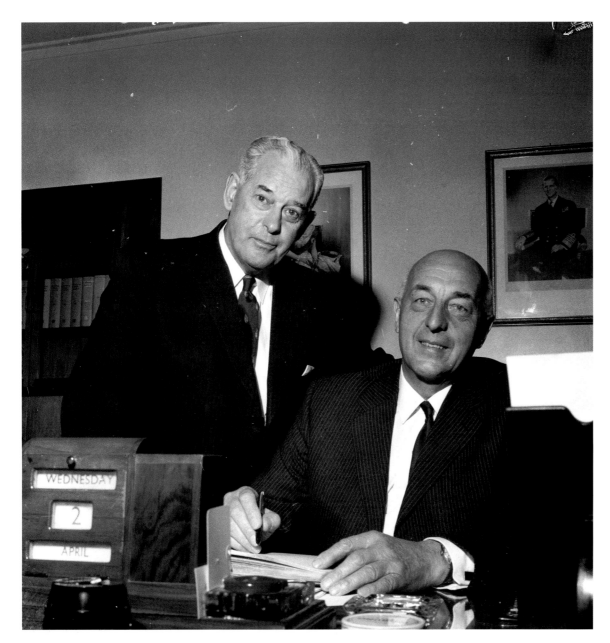

Prime Minister Keith Holyoake (left) had accepted British membership of the European Economic Community was inevitable and called the National Development Conference with economic sectors to report in 1969. In that year, vice-president of the EEC, Dr Sicco Mansholt (right), visited New Zealand and had discussions with the Cabinet economic committee.

1882, British interests took a large measure of control over meat processing plants, shipping services and the marketing of meat and dairy produce. In 1901 Trade Commissioner Graham Gow, appointed by the new Department of Industries and Commerce, travelled widely within the country to acquire the information and 'samples' for selling missions he was to undertake to South Africa, India, China, Japan and elsewhere. The department aimed to expand overseas markets for New Zealand meat and dairy produce. Documents suggest that both farming organisations and governments found the freezing companies, at least in the North Island, were resisting attempts to diversify markets for meat and to branch out from the export of bulk carcasses to a single outlet, Smithfield in London.

The department reported a successful export sample of New Zealand lamb to New York and claimed a great mistake was being made by freezing companies in confining meat exports to a bulk trade to London – to 'secure a substantial profit for themselves' – when American, Spanish and Scandinavian markets were inviting business. Trade Commissioner Gow, by 1902 in Britain, insisted unexploited markets existed in many other parts of the United Kingdom where companies were ready to take wholesale quantities of lamb. The department also insisted that the freezing companies in the North Island had adopted an unfair fixed-price procedure for purchase of fat stock from farmers. His report criticised the 'Home dealers' at Smithfield for opposing branding of the New Zealand product because they would be robbed of the immense profits that could be made out of the meat by selling it as 'home-grown' (and thus not frozen). Branding was a problem that continued over a number of years in more than one market. Australian wholesale purchasers of New Zealand butter in 1902 were reported to be removing the original brands and substituting others.

Gow reported visits to Aberdeen, Inverness and Dundee where he found limited supplies for sale of 'highly regarded' New Zealand butter, cheese and lamb. He thought the best way to expand trade to Scotland was through Manchester, adding, 'I am of the opinion that steps should be taken at once to make vigorous and sustained efforts to open a market in the Manchester district and the North of England generally.' He even suggested, prescient by half a century, that 'a trade in wood-pulp might be opened up'. However, the following year he reported that after a 'careful investigation as to prices, etc,' he considered it would not be wise to start a wood-pulp industry because the manufacture required 'very expensive machinery' and the country could not compete with Norway and Canada.

Gow had all the overdrive of a salesperson. Part of

The freezing industry

Once the potential for frozen meat exports to Britain was recognised in 1882, processing plants for the killing and freezing of mutton and lamb sprang up around the country. Among the first works were those built by the New Zealand Refrigerating Company (at Burnham, near Christchurch), the Gear Meat Company (in the Hutt Valley), the Wellington Meat Export Company (at Ngauranga on the Hutt Road), and the Nelson Brothers Ltd (near Hastings). By 1885 an astonishing forty-five freezing works had been built around the country, and in the regions where sheep farming predominated (with ancillary beef production) the processing plants expanded and proliferated.

When the industry was delicensed (and thus deregulated) in 1980 most of the freezing works were huge, old and inefficient. They were often situated at railheads near cities and provincial towns to tap labour pools or, preferably, near ports. In Wellington, a major plant was at Ngauranga, between the city and suburbs such as Khandallah and Johnsonville, on the road to the Hutt Valley, and another was close to the foreshore at Petone. As the cities grew, the smell and effluent became problems.

Before deregulation the companies did not have much incentive to strive for efficiency. The EEC, the United States and other customer nations insisted on certain standards of hygiene, but the companies –

his brief was to tout for tourism and he wrote that in the North of England 'several gentlemen have announced to me their intention of proceeding to New Zealand at an early date for a holiday tour'. He next moved to South Africa from where he sent a huge report on the possibilities of setting up a shipping service between the two countries to foster a frozen meat trade he was sure would develop into a mainstream business.

The following year the indefatigable Gow visited Japan, China, India, Ceylon and Singapore and was enthusiastic for trade prospects with Japan which already had shipping services to Australian ports and, he thought, could easily be persuaded to continue on to New Zealand. Little progress was made but the government persisted with efforts to diversify, despite tariffs. In 1914 the Department of Agriculture, Industries and Commerce reported butter exports to Vancouver of 60,000 hundredweight in the previous year after a period of 'surprising' growth.

For the first twenty-five or thirty years of refrigerated exports, the concerns were getting consistency of product quality from the many freezing works

only about one-third of them producer co-operatives – could achieve acceptable procedures without actually modernising the way in which they operated. Hygiene was the guiding principle not plant efficiency.

Many of the workers, chain butchers for example, needed great strength and stamina and were highly paid on throughput rates. The drive for automation began not long before delicensing and threw many butchers out of work. The deregulation of the farming sector in the mid-1980s exacerbated these unemployment rates and led to a proliferation of smaller processing plants. Meanwhile, subsidies to sheep farmers, such as supplementary minimum payments (SMPs), had so distorted the relationship between sheep numbers and plant capacity that the industry was in disarray.

During the 1970s and early 1980s farmers wanted a higher stake in the processing industry, and during the crises of the 1980s the Meat Producers' Board decided it would reorganise ownership and invested

By the 1980s freezing-works plants were generally old and inefficient with huge cooling rooms like the one pictured left. But new plants were being built, like the one pictured at Moerewa in the Bay of Islands in 1993. They were hygienic and at least partly automated. Moerewa is capable of handling a lamb kill of seven thousand a week and a beef kill of three thousand. It is also certified as a halal slaughtering plant.

money into buying shares in meat companies. The board, which became a non-trading producer board under the Meat Board Act 1997, received revenue from investments, including those in the industry, and from fees it levied on the slaughter of sheep, cattle and goats. By the 1990s three producer-owned organisations owned more than two-thirds of the processing operations: Alliance, PPCS and Affco. It was believed that when they took over from the British-owned processing companies cost savings would result in higher returns for farmers, but this proved difficult even in an industry turning over more than four billion dollars a year.

The dangers of economic dependence were not unknown during the first third of the twentieth century but no one was seriously attending to them. In the period immediately before World War One, the limitations of the British market became apparent. The confrontation between farmers and urban workers in 1912 and 1913 was not just ideological, as William Massey's government insisted. Unemployment was high and growing, and workers' wages had slipped behind the incomes of farmers and the self-employed. The war and the boom immediately afterwards changed that. But a series of recessions during the 1920s prompted some observers to reconsider the continuing lassitude of New Zealand's approach to marketing. The recession in 1921 was so sharp it was a factor in forcing rehabilitation soldier/farmers off their recently acquired land.

Most extraordinary, given all these economic indicators, was New Zealand's continued ignorance of the British market and its consumers. D. O. Williams, a lecturer in agricultural economics at Massey Agricultural College, wrote fifteen years later, in 1936:

> A definite obscurity prevails as to the particular requirements of our customers . . . At present we labour in much ignorance and do little to mend our ignorance. But if we intend seriously to find some escape from an intractable quantity position, we must convert ourselves to a religion of quality and of market specialisation, and this calls for a knowledge of our customer's requirements far more detailed and accurate than anything we possess at present.

The problem associated with the absence of hard-headed marketing began with the trust the colonials had for the people at 'Home'. This assurance may not have been quite as exaggerated under the Liberal administrations (1890–1911) but Prime Minister William Massey, who presided over a government of farmers, was an English-born Anglophile who went to Mother England as often as he could while in office. He was a very able politician but abject in his attitude towards Britain. Consequently, not much progress was made.

Prime Minister from 1912–25, William Massey was an English-born Anglophile, an able politician but abject in his attitude towards Britain.

and dairy factories that sprang up around the country. To this end the government of the day brought out experts from Britain and installed meat and dairy inspectors at the factories and in London where distributors had most control. But fraudulent abuse of brands continued for many years. It wasn't until twenty years after the refrigerated trade commenced that New Zealand began thinking about cold storage in London to manipulate the timing of when produce went onto the market.

As the Depression of the 1930s was coming to its end, New Zealand was in the mood for self-examination, even self-flagellation. The country was still so totally dependent on Britain and its consumers that a report at the time said: 'More urgent, perhaps, than encouraging a greater diversification of our exports, is the need to secure additional markets. To a dangerous extent our exports are concentrated in one market.' More than eighty per cent of New Zealand's exported butter went to Britain, eighty per cent of the beef and fifty per cent of the wool, which was the only export commodity with an international market. Markets outside Britain had diminished during the Depression, and the country remained exposed to every British price crisis. An additional worry was that the English farmer applied pressure to limit New Zealand competition through tariffs and quotas.

Agriculture was also critically important to the functioning of the domestic economy. In the 1920s, before the big Depression struck, nearly seventy per cent of the total national value of production was primary products. And these products represented ninety-four per cent of total external trade. The occupation of one-quarter of all 'breadwinners' in the country directly involved the production of food, and a government report at the time said:

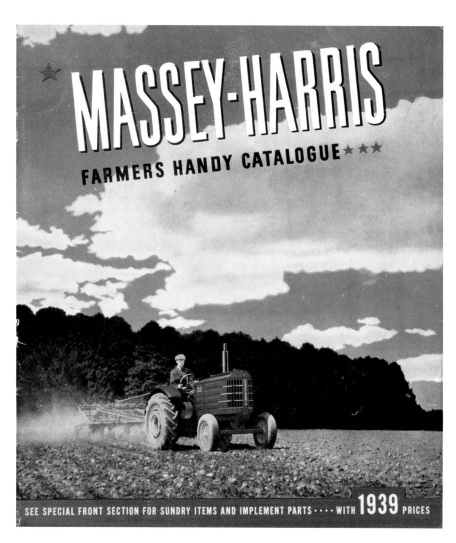

> The pervasiveness of farming in our national life is reflected in other ways not capable of numerical statement. The organisation of New Zealand on a specialised farming basis, producing a limited range of major products to be marketed mainly overseas and dependent on large importations for its secondary requirements and on large borrowings for its capital need is the explanation of many of our most pressing problems . . . Even though the prices of any or all of his products fall catastrophically, as they have, the farmer feels that he has no option but to continue producing. Moreover his external costs, as well as import prices, are much more rigid than his receipts from overseas selling . . .

The truth was that New Zealand had always had limited resources beyond agriculture, had a small population, and therefore relied almost totally for economic health on exports of a range of products that the British, almost alone, wanted. At that time, Britain bought ninety-four per cent of the lamb and mutton that was traded internationally, about eighty per cent of the beef, about half the cheese (and the cheddar market was only a small wedge of that); and Britain and Germany bought ninety per cent of the butter traded internationally – so where else was there to sell our produce?

During World War Two and in the immediate post-war years New Zealand governments and producer agencies continued the bulk-selling arrangements. In 1943 a Dairy Industry Commission decided 'the opening of new markets to dispose of surplus production can best be undertaken by a single central body acting in the national interest and having due regard to all branches of production'. The result was contracts with the United Kingdom Ministry of Food that went through until 1954. New Zealand gained the right to sell up to fifteen per cent of butter and cheese to markets outside Britain in 1951, and the Department of Agriculture reported in 1952 that prices in other markets were higher – in fact, brought in £1.3 million more than had the produce been sold in Britain. And yet these sales were cut back to five per cent the following year when Britain appealed for more food. Similar bulk-purchase contracts with Britain covered meat exports after the war. The Department of Agriculture reported: 'While the meat shortage continues in the United Kingdom only very small quantities are being shipped elsewhere, but the possibility of exploring other markets, especially the North American market, is receiving consideration.'

In 1964 – midway between the end of bulk purchasing in 1954 and Britain joining Europe in 1973 – a seminar was held at Waikato University on 'Diversification of Primary Industry'. (Some background: the British Government that year subsidised its agriculture with about £360 million in direct farmer subsidies that worked this way: New Zealand lamb had free access at the ruling market price which was twenty-seven pence per pound but the British farmer had his return built up to an agreed thirty-nine pence per pound.) One contributor was Dr A. H. Kirton of the Ruakura Agricultural Research Centre. Kirton started by noting that the targets committee of the Agricultural Development Conference, set up by the government the year before, had found 'that we will have to increase our animal population by four per cent annually in order to maintain [our] export earnings and present standard of living'. He then asked where producers would sell this additional lamb and mutton, and followed with a long lamentation on the lack of knowledge available about primary industries in general but the meat industry especially. He noted that the application of scientific principles of breeding in sheep and beef cattle was forty years behind the dairy industry.

Kirton said agriculture required an efficient market research and sales organisation. Better 'factual' information was needed 'on the quantities, quality and suitability of our present meat products even in our main markets'. 'What proportions,' he asked, 'of the various grades of lamb carcasses are sold in the various regions of the United Kingdom throughout the year? At what level of fatness does the British butcher regard a lamb as overfat, and at what level do the Americans regard the lamb as overfat?' He mentioned also the lack of specific knowledge on access to new markets.

The seminar noted, however, that progress had been made since 1954. Meat exports then earned £59 million, eighty-seven per cent of it from the United Kingdom. In the 1963–64 season earnings from lamb, mutton and beef and all by-products were a record £119 million, only fifty-three per cent of it from the United Kingdom. But this change was almost entirely due to the sudden opening of the American market to manufacturing beef and some access to Japan. Both these markets had variable restrictions which could change from year to year.

Those involved in the seminar were either people from the industry or those deeply involved in industry studies at universities; so the vision, Kirton's apart, was generally narrow with emphasis on the difficulties of diversification rather than on marketing needs. The dairy contribution stressed the difficulty of finding volume markets for cheese types other than cheddar and suggested that product diversification depended entirely on adequate prices for butter. At that time, butter earned £56 million, cheese £21 million and all other manufactured dairy products £12 million.

The first British gestures to Europe in the early 1960s were rebuffed but the inevitability of membership was becoming apparent. New Zealand negotiated a reasonable phase-out deal with Britain for when it did join, with gradually diminishing quotas. One advantage the country had in these talks that continued through the late 1960s and early 1970s was the Minister of Overseas Trade in Holyoake administrations from 1960 through to 1972 – Jack Marshall. He was also Deputy Prime Minister through that whole period. He became Prime Minister briefly before being dethroned by Rob Muldoon.

By the 1950s farmers and governments were at last beginning to accept that the future lay with exploring what consumers wanted, especially in North America and Asia, within the unfortunate boundaries of tariffs and quotas. So in the 1965–66

Ten years before Britain joined Europe, Waikato scientist Dr A. H. Kirton insisted New Zealand agriculture needed an efficient market research and sales organisation to find better information on the suitability of products even in the main markets.

season, well over half the beef and mutton exported went to Japan and the United States, although almost all the lamb continued to go to Britain. With the traditional market in jeopardy because of the EEC, the dairy industry was also becoming more nimble.

An example of how New Zealanders were starting to rise from their long-time marketing slumber was a serious examination in the mid-1960s of the potential for bulk air-freighting export meat, especially to Asia. Food was already being stowed as filler-freight on scheduled passenger jet services but only for luxury markets because of the cost. However, the 'jumbo' Boeing 747 was on the horizon, providing a huge leap in capacity and, concomitantly, lower costs. The meat industry seriously investigated the possibility of using a fleet of jumbo freighters to fly large consignments of chilled and frozen cuts to an already diversifying market. They even examined the concept of constructing a special airport – Rangitikei was one site – as a central take-off point for worldwide distribution of produce. It was a brave concept but the economics did not stack up.

In summary, for near enough to a century, the farming of New Zealand consisted of producing large quantities of a narrow range of low-price

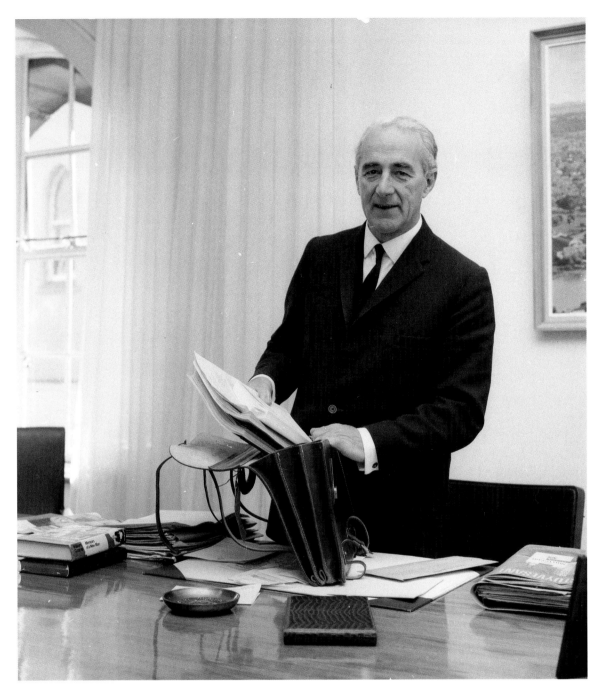

One advantage New Zealand had in negotiating interim arrangements for this country's produce as Britain prepared to join the European Common Market was that Jack Marshall was the Minister of Overseas Trade throughout the negotiating period, for more than ten years.

commodities to satisfy the notoriously undiscerning palates of the English. It was a gloriously simple business, requiring no entrepreneurial or marketing skills. If business went bad occasionally, it was because the world broke stride in a cyclical slump and everyone suffered together, not because Britain had decided to shop elsewhere.

The cheap food policy that had been a basic tenet of British political policy for many years was a major reason why many Britons baulked at European Economic Community membership in the early post-war period; so when the UK–EEC marriage occurred, New Zealand at first saw the move as apocalyptic. She had been wholesaler for an assured retail market, sedate supplier without serious competition, the mistress of the British food business; and she had lived better than the wife, the British farmer, despite a period in the 1930s when attempts were made to impose quotas.

Well, the kept lady was going to have to take to the streets and hustle while uttering airily such incantations as 'diversification' and 'restructuring'. Gradually the lady came to understand that her goods were attractive to a wider range of people than she had thought; that her favours were more frequently and widely sought than she had expected. She could at last flaunt her wares with confidence and skill.

CHAPTER THREE

FOOD AND THE SOCIAL REVOLUTION

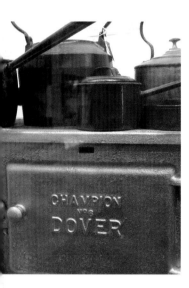

Above: New Zealand cooking was at its best when a dish did not need hovering over, could be put inside an oven or a pot, and left pretty much to its own devices. The Champion range was a favourite.

Opposite: The symbol of change was wine. Sherries and ports and plonk were sold in half-gallon flagons, much as beer was. Pictured is a half-gallon filler in 1960.

THE SWIFTEST social revolution in New Zealand history had a false start in 1962 when a tiny dairy company in Opouriao Valley near Whakatane decided to produce a batch of camembert. A Dutch immigrant on the staff, Peter Assink, was called upon to make the famous French cheese. He did so from a book of instructions, and sought the advice of a local Swiss whose knowledge was also 'theoretical'.

Assink had no experience of making cheese in Holland. He arrived in New Zealand in 1949 at the age of twenty-two, married, and worked on dairy farms in the Bay of Plenty before joining the staff of the Opouriao Dairy Company. The company was founded in 1900 after frosts had burnt black the valley's staple maize crop and milking cows seemed to the locals to be a less risky if more arduous alternative to cropping.

No one seems to remember what inspired the dairy company management sixty years later to pioneer a cheese as removed from real New Zealand life as camembert was at that time. That first year, though, Assink and his team got the moisture content wrong and the factory had a large batch of green cheese on its hands – which it buried.

Undaunted, the following year they made not only passable camembert but brie (also French) and gruyère (Swiss) as well. The small company also produced processed cheeses and spreads, using additives such as ham, bacon and cumin seed. The European-style cheeses didn't sell in the Bay of Plenty so an agent in Auckland was employed. Sales were negligible but the agent was encouraged. Alas, the next year, 1964, Opouriao was merged with the Rangitaiki Plains Dairy Company. Exotic cheeses were abandoned as the factory concentrated on mainstream casein production. Assink lost his job and Opouriao Valley returned to the calmer, cheddar-chomping cultural times in which the rest of the country had remained. The Auckland agent was intrigued enough to hire Assink and take him to Auckland with the idea of continuing with camembert. It never happened. Said Assink years later, 'Kiwis at the time didn't know what camembert was. They hardly knew what wine was.'

This national ignorance did not last much longer though. The Opouriao Dairy Company's only mistake was to be ahead of its time. New Zealand's great leap forward from bland social conformity to a modern, expansive society was stalled for only about another decade. For while farmers and governments were struggling to adjust to the twin trials of Britain joining Europe and the deregulation of the economy, the New Zealand community – both rural and urban – was transforming itself.

The symbol of the change was wine. The switch from sherries and ports to table wines came quickly. New Zealand whites were immediately appealing, if underdeveloped, by the late 1970s and local reds gradually asserted themselves against imports from Australia where experience in making them was deeper. By the 1990s New Zealanders were making, drinking and exporting wines such as sauvignon blanc and pinot noir that were as good as or better than any produced in countries with hundreds of years of wine-making experience.

Overcooked cabbage was a symbol of the way green vegetables were generally treated in years gone by, and it was one of the favourite three veg eaten with meat before the revolution.

Farming and food have always interacted within any society in a kind of circular way. People ate what their farmers made available locally and farmers produced what people seemed to want. It took a political and social revolution to interrupt that rhythm. New Zealand had, from the beginning of European settlement, reproduced Britain's pastoral farming commodities, grown typical English vegetables in their gardens, and adopted British food and drink styles as well as their understated, conformist clothing and mode of living. This way of life was marooned atavistically in New Zealand for the best part of a century while Britain suffered the continuing social upheavals of the Industrial Revolution that suddenly, painfully, urbanised its populace. It may have been somewhat dull in New Zealand but it was comfortable, and plain food repetitively served was not confined to this country.

Because Dunedin was the wealthiest city from the earliest days, the Scottish influence was profound and to a large degree it calibrated colonial life. A year after the Indian Mutiny in 1857, Victorian England's pre-eminent arts and social commentator John Ruskin tried to understand why Indians, with their great artistic tradition, could be so barbaric compared with the Scots. He said he had been to the north of Scotland the previous summer and felt 'a peculiar painfulness in its scenery, caused by the non-manifestation of the powers of human art'. It was the first time in his life he had been in any country seemingly possessing no valuable monuments or examples of art. Whereas, among the models of art, according to Ruskin:

> Out of the peat cottages come faith, courage, self-sacrifice, purity and piety, and whatever else is fruitful in the work of Heaven; out of the ivory palace come treachery, cruelty, cowardice, idolatry, bestiality, – whatever else is fruitful in the work of Hell.

Although this moralistic tirade was prompted by British ignorance and arrogance, it did give a contemporary picture of the ascetic, moralistic Scots who had settled Otago ten years before.

New Zealanders had from the beginning been strikingly progressive, confident, inventive and hands-on when it came to the practical business of making things work more efficiently on the farm; but visitor after visitor had noted their extreme social conservatism, bred by extreme isolation. Only the wealthy could afford to travel back to what they called 'Home' once they had arrived in New Zealand, except for servicemen who saw the narrowed vision of only death and turmoil on their travels.

The breeze of change, faint at first, came in the late 1960s, after the first wave of young Kiwis returned from their OE in Britain, and after television had ushered in images of the wider world and jet aircraft began to bring in tourists. The country was ten years behind the same cultural revolution in Australia where it had been enhanced immediately after the war by shiploads of Italian and Greek migrants.

Little wonder, then, that New Zealand culinary standards were plain and had hardly moved from the British middle-class styles of a century earlier. In 1953 English writer Eric Linklater wrote of a visit to New Zealand in his *A Year of Space: A Chapter in Autobiography*:

> But New Zealanders, like the Scots, think that baking is the better part of cookery, and spend their ingenuity, exhaust their interest, on cakes and pastries and ebullient, vast cream sponges. Soup is neglected, meat mishandled. I have seen their admirable mutton brought upon the table in such miserable shape that the hogget – so they call a sheep of uncertain age – appeared to have been killed by a bomb, and the fragments of its carcase incinerated in the resultant fire.

Another more vituperative departing broadside at New Zealand's pitiable cuisine came from eminent British media person, BBC broadcaster Wynford Vaughan-Thomas, three years after Linklater's book. In *Farewell to New Zealand* he wrote:

> Super-suburbia of the Southern Seas,
> Nature's – and Reason's – true Antipodes,
> Hail, dauntless pioneers, intrepid souls,
> Who cleared the Bush – to make a lawn for bowls . . .
> Saved by the Wowsers from the Devil's tricks,
> Your shops, your pubs, your minds all close at six.
> The Wharfies' Heaven, the gourmets Purga'try:
> Ice cream on mutton, swilled around in tea!
> A Maori fisherman, the legends say,
> Dredged up New Zealand in a single day.
> I've seen the catch, and here's my parting crack –
> It's under-sized; for God's sake throw it back!

At that time, such withering ripostes hurt New Zealanders who, however, found later they didn't care overmuch about the opinions of visiting Brits. They had been there and seen them in their habitat and knew that they didn't have much to envy. And, even more extraordinary, in the twenty-first century the average Kiwi cuisine at home and in restaurants was better than theirs, and some of the wine indeed much better than France's, although French cuisine still led the way. Both Linklater and Vaughan-Thomas were comparing ordinary fare in New Zealand's relatively small towns with restaurants in London and other large-population cities in their day.

Furthermore, as any Kiwi who was around in 1953 and for a few years afterwards will know, hogget is the most delectable eating stage in the life of a sheep – its age not indeterminate, as Linklater had suggested, but approaching one year, just before the two-tooth stage – with the tenderness of a lamb but a deeper flavour. Exquisite, if roasted by a cook of experience, with potatoes, pumpkin, parsnips, onions and kumara around it – a practice pretty well confined to New Zealand. Elsewhere, cooks roasted meat on its own. Linklater must have been in a rare household or restaurant because the one meal New Zealand women, and especially country women, could cook, was roast lamb, beef or pork.

However, anyone old enough to have lived through the 1960s would have to concede that the delicacy of cooking in New Zealand at the time was not high and was at its best when a dish did not need hovering over, could be put inside an oven or a pot and left pretty much to its own devices – namely, roast meat, steak and kidney pie, and stews. Cooking was a lowly trade outside the home, and an arduous one inside it. It was almost entirely the work of women. The only places men were trained to cook were at sea, in the armed forces or in jail. Training in cooking that included foreign dishes did not become generally available and popular until the 1970s. The social difference between a twenty-first-century chef and an early twentieth-century cook is the difference between a lord and a peasant, the status of one inflated and of the other unfair. The roast apart, Linklater was, however, correct that meat was generally mishandled. Here is some advice on 'The Art of Grilling' from the *New Zealand Farmer and Stock and Station Journal* of April 1930:

> It is an art – the cooking of chops and steaks, and even of making toast, under the griller of the gas stove. But it is one that is very quickly mastered, although the young amateur cook always seems diffident about attempting it. There are two things to remember . . . the grill itself must be red hot before it is used, and so must the bars of the gridiron. The second is that the food must be watched all the time it is grilling. . . Grease the bars of the grid thoroughly as soon as the grill is red hot, and put the steak on them. Push it into place quickly and lower the gas slightly if the flames threaten to touch the meat. As soon as the outside has been sealed by the heat – it takes from two to three minutes – turn the gas down by half and leave it like that for five or six minutes.

Steak tended therefore to be eschewed because it could not be chewed. The range of cooking advice in magazines such as the *New Zealand Farmer* was narrow, except for cakes, biscuits and puddings. Linklater was also right about the predilection of women, again especially country women, for baking rather than cooking. *Aunt Daisy's Ultimate Cookery Book*, published in 1959, contained 1300 recipes in 298 pages – 201 of them dedicated to sauces, puddings, cakes and other sweets and desserts. It was no accident that New Zealand's all-time bestselling cookbook is sponsored by Edmonds, manufacturer of baking powder. The way lives were lived then had a bearing on this predilection. Household meals were simply enough cooked but baking required higher skills, more measuring, better timing and more supervision.

Restaurants were rare, mainly in posh hotels, expensive because labour costs were high, and patronised by diners only on very special occasions, such as weddings and twenty-first birthdays. New Zealand, by the nature of its widely spread population in suburbs and towns, did not possess a restaurant culture. Apart from a lack of discretionary income in most families, people simply didn't have the mobility to move around freely in the evening. City workers went home at five o'clock from work, or six o'clock if they went to the pub, had an early meal, spent the rest of the evening at home and went to bed. Farmers also tended to eat early, listen to the wireless perhaps, and go to bed early in order to rise early. When functions were held in the country they were seldom sit-down dinners but rather ladies-bring-a-plate occasions – meetings, 'socials', concerts or parties. When people called at the weekend, they tended to do so for morning or afternoon tea, or supper – another occasion for scones, tall fluffy sponges and large lavish cakes, usually embellished with whipped cream.

The revolution that transformed the New Zealand way of living began in earnest with a referendum in 1967 that ended fifty years of six o'clock closing of hotel bars – eighteen years after such a move had been resoundingly beaten in a similar referendum. Moves had been made earlier in the 1960s to allow some restaurants to serve alcohol with meals, but under restrictive conditions. When

Few people in New Zealand used strong spices of any sort until the food revolution of the 1970s. Virtually no one used garlic in cooking and the smell of it on anyone's breath was regarded as offensive. By the end of the twentieth century it had become a standard ingredient not only in ethnic dishes but in the cooking of local food as well. 'Harry' and workers are harvesting garlic on his Whangarei Heads farm, Northland, but the local industry is coming under pressure from low-priced Chinese imports.

it was realised in the years after 1967 that drunkenness and debauchery would not break out in public with extended pub opening hours, the door creakily opened further and allowed gradual extensions of licensing regulations that eventually extended to cafés and corner stores, conditions as loosely applied as anywhere in the world.

Perhaps the first distinct signs of new culinary sensibilities came in the forms of olive oil and garlic, both ingredients which New Zealanders had, without even trying them, regarded with anathema for cultural reasons. When the *Woman's Weekly* began to advocate the use of these two ingredients in its cooking columns it attracted blistering letters of complaint. Looking back, one can remember the dish that converted the home cook to the new cuisine as the shrimp cocktail – ubiquitous until it became an offending cliché.

But the change was not, as it is often portrayed, the result of a nation emerging from studied ignorance. The cause was the convergence of a number of disparate factors: the rapid spread of the home refrigerator-freezer; the proliferation of other appliances, such as washing machines, freeing housewives from much hard labour; improved international and national transport systems, and packaging, reducing dependence on local produce and on seasonality; the end of, first, rationing and then the loosening of the import licensing system; foreign tourists arriving in numbers; young Kiwis returning from their London OE; an expanding amount of income for discretionary spending in what was then one of the two or three wealthiest countries in the world.

Ridicule of pre-1970s New Zealand cuisine by food commentators since the 1990s has become a national confessional – as though Kiwis were expiating some unique sin, as if they confined themselves to a narrow range of food from lack of knowledge, whereas good reasons explain it in retrospect. Housewives, and especially rural women, cooked in coal-range ovens, sometimes even camp ovens, and needed food that could be left because the technology did not allow for fine-tuning, and because they were so unceasingly busy with other diurnal chores around the house or on the farm at a time when families were bigger and before household appliances became common. Pioneer teacher and home science specialist Tui Flower recalls that an American firm claimed from research that a woman washing the heavy, coarse-fibred clothes of the time, using a copper, tub and hand-wringer, was expending the same amount of energy as a woman swimming breaststroke for eight kilometres. In addition, the cooking cycle for meals had to take into consideration the keeping qualities of meat, milk and other ingredients before refrigeration became widespread. Milk puddings and other dishes, for example, were common not only because they could be left to cook but in order to use up milk before it turned sour. Not just in New Zealand, but all over the world, people ate within a fairly narrow range of food, dictated by the availability of ingredients. Baking, on the other hand, was a craft that required care and attention and for the exercise of which women cleared time, say, once a week.

Seasonality in a food-producing country also had its charms. People waited for the few months in the year in which some fruits and vegetables became

The first distinct signs of new ingredients arrived with olive oil and garlic, once seldom if ever used in New Zealand cuisine. But now olive plantations are becoming increasingly common in the landscape, like the one pictured in Nelson.

Before the days of appliances, a woman's diurnal chores were estimated as the equivalent of swimming breaststroke for eight kilometres, and a rural woman generally had additional chores.

available and relished them in their freshness; whereas some fruits, available right around the year now, imported from overseas, are expensive, lack freshness and flavour, and quickly go bad.

Nevertheless, this much reviled Kiwi diet was nutritious enough for observers during World War Two to remark that while New Zealand soldiers were not tall they were strong and stocky and had marked resources of stamina. Perhaps the reason was that, even if the cooking was not delicately done, food was abundant in New Zealand, even relatively so during the Depression, and milk, meat and potatoes were inexpensive.

The standard dinner until at least the 1960s was meat with potatoes and two or three, usually overcooked, greens. These meals reflected what was available. Lamb, beef and especially offal meats were cheap and plentiful and those vegetables easily grown in home gardens were fresh and available seasonally. Potatoes, pumpkins, onions and turnips, and to a lesser extent carrots and parsnips, could be stored and were therefore available through much of the year, but marrows and green vegetables – cabbages (two varieties), lettuces (one variety), silver beet, dwarf (French) beans, scarlet runners, broad beans, green peas – had short lives out of the ground and could be ready for picking for no more than three months in most parts of the country. Broccoli, asparagus, avocados and courgettes were rare, although courgettes matured as marrows were common.

Potatoes were eaten with every main meal and seasonality had an effect on them too. 'New' potatoes – small, white and freshly dug – went with lamb and green peas (with mint) for many a Christmas dinner. During most of the year, 'old' or 'main crop' potatoes were eaten, simply boiled and sometimes mashed, or baked in their jackets, or roasted with a joint of meat. Every home gardener grew potatoes, some enough to get through the

Potatoes were eaten with every main meal. A seventy-five-year-old woman helps harvest a crop on a small farm in Canterbury in 1948.

year, although they were so staple it required a lot of garden space to grow that many for a family.

Meat was the other staple that most main meals were constructed around. Large joints of beef and lamb and big pots of boiled corned beef or mutton were cooked hot for a main meal. What was left over was sliced and eaten as cold meat with vegetables the following night, and the remainder minced and reused for shepherd's (or cottage) pie, or possibly savoury mince on the third night. In the days before refrigerators this was prudent use of meat. Fried (rarely grilled) lamb and pork chops, pork sausages and offal meats, accompanied by the usual vegetables were the fill-ins. Dripping was used for cooking and the intake of animal fats was very high. For example, dripping would be placed around roasts, which were themselves usually surrounded by their own fat. After the meat was cooked, the dripping left in the roasting dish was poured into a fat tin and kept for frying and the next roast. The effects of so much fat in the diet were significantly reduced by walking and other activity concomitant with daily work in the days before motor cars, farm-bikes and wholly automated machinery. People of the time would have been enormously amused to be told that fifteen minutes' walking twice a week would reduce their susceptibility to dementia. Oh, that they could have been so lucky!

Offal meats were also basic and vigorously advocated by nutritionists for their food value. One of the most influential, Dr Guy Chapman, said: 'and remember that liver, kidney, tripe, heart and so on are really the most valuable meats that you can eat and, no matter how much money you have, they should be included at least once weekly, and better still, twice, in your diet.' Chapman even advocated using liver, kidney or heart twice a week to help prevent constipation, which was a common, indeed pervasive, complaint of the time and one on which he devoted many a radio broadcast.

Following pages: Along with potatoes, pumpkins and onions, turnips could be stored and available through much of the year. Almost never seen now, swedes were not only used as stock feed but as a staple vegetable in the home. Dairy farmers had to be careful about when and how much swede cows could eat because too many could taint milk.

Seasonality for fruit and vegetables does not exist in supermarkets any more but a demand for freshly picked or dug fruit and vegetables is reflected in the proliferation of local farmers' markets, like this one in Dunedin.

Sweetbreads, black and white puddings and brains were also regularly eaten, along with ox and lamb tongue and pig trotters. Offal is no longer easily available or cheap and has had a bad press from the medical profession in recent years because of its generally high cholesterol content, although its high levels of iron have been missed by many. The dramatic break from the past is illustrated by the fact that offal foods, once so commonly on display, were gradually removed from butcher's shop windows because the sight of them offended young shoppers.

Fish was a Friday thing, an imperative for Catholics but also a chance for a change of diet for the rest. Fish was not as common in meals then as it was to become later, although fish-and-chip shops abounded. Fish shops also sold poultry and rabbit. Rabbits were widely eaten – as much or more than poultry. Mainly bred and farmed for eggs, poultry was seldom used as a meat, and bore no resemblance to the specially bred, pumped-up birds common since the advent of supermarkets.

My own memories of food before the 1960s are that it was certainly plentiful. Breakfasts were hearty, with cereal (usually porridge), toast and sometimes an egg. School lunches were rather skimpy by today's standards but dinner was always plentiful and certainly rich in carbohydrate and protein. In smaller towns and cities in early years, families would gather for dinner in the middle of the day. Schools were often in the close neighbourhood, and along with fathers' work a short walk or bicycle ride away from home. Later, as cities grew bigger, dinner at midday was impracticable, so it was shifted to as soon after 5 p.m. as father could get home from work – or after 6 p.m. if he went to the pub. Eating after 6 p.m. was considered a late meal. Three meals a day – with breakfast around 7.30 a.m., lunch around noon, and dinner around six – was the standard.

At secondary school, my lunches consisted almost invariably of lettuce, cheese and Marmite or Vegemite sandwiches with biscuits or a piece of cake and a piece of fruit; and rare among my contemporaries, I loved the free school milk. White bread was the norm – indeed, brown bread was almost entirely white bread dyed. Did I get sick of my sandwiches? Well, we kids often swapped them among ourselves to get variety – something that would horrify mothers of the twenty-first century – but I was always hungry, so getting sick of them was not an issue. Often enough today I make myself a lettuce, cheese and Vegemite sandwich, but with wholemeal bread, and still relish it. And I still cook what I call trad meals – meat or fish and three or four vegetables – and enjoy them too.

CHAPTER FOUR

OPENING THE WORLD TO NEW ZEALAND MILK

The dairy industry, fully recovered from the pessimism that shrouded the future in the 1970s and 1980s, entered the twenty-first century shaping itself for leadership in global export markets. Cows numbered 3.9 million in 12,000 herds producing 14.8 billion litres of milk, 95 per cent of which was exported in one processed form or another: milk powders, cream products (butter, etc.), cheeses, and milk and whey proteins. These products were New Zealand's biggest single export earners.

International prices would continue to rise and fall, subject to consumer demand and to the distortions of subsidies to dairying in developed countries, but optimism was abroad as farmers invested in their industry and new farms crept beyond the fringes of traditional dairying country. But for all that, more than eighty per cent of dairy cows were grazing the lush pastures of the North Island – more than a third of them (1.6 million) in the South Auckland-Waikato-Bay of Plenty region. Taranaki was next with 530,000. In the South Island, though, Canterbury (452,000) and Southland (357,000), where land was cheaper, were advancing.

It had been a long, hard haul through such crises as Britain joining the European Economic Community in 1973 and the deregulation of the New Zealand economy in the following decade. Soon after World War Two, butter and cheddar cheese export sales began to contract. Britain traditionally absorbed more than eighty per cent of the world butter trade but demand in that market was stagnant as the fifteen-year-old New Zealand bulk export agreement with Britain expired in 1954. About the same time, the international market was being corrupted as European countries expanded dairy production, supported by subsidies, from 830,000 tonnes to 930,000 between 1954 and 1957, the year in which the European Economic Community was formed. Europe was recovering from the war, and under the EEC regime retreated behind import barriers. Next came the spectre of Britain also disappearing behind this European protectionist fence and stimulating even greater overproduction. All the while, the United States was dumping its subsidy-promoted surpluses on world markets, sending international prices into a nosedive. Against this calamitous backdrop, the output of New Zealand butter and cheddar cheese was growing.

Wanted were new buyers in new countries for milk products. The Dairy Board, responsible for international marketing, shrewdly dispatched technical staff into Asian, Middle Eastern and Latin American countries to set up recombining plants in which products with local demand could be manufactured from imported New Zealand milk powder. The board had acquired some expertise in manufacturing offshore. A recombining plant had been set up in the West Indies in the 1950s to meet local demand, mainly because that was where cargo ships from New Zealand stopped for coaling on their way to Britain. It was a small operation but successful and had furnished technical staff with knowledge and experience that proved useful a

By 1975, as the average herd size grew from sixty-five to around a hundred and as roads improved, these early milk tankers helped factories become more centralised and bigger.

decade later. A plant was opened in Singapore in 1961 followed by others in South-East Asia and the Mauritius. Skim-milk powder and casein, by-products of butter-making, became major exports, adding nutrition to stock feed, and also used in the manufacture of sweets and beverages, bread and other bakery products, and even sausages. In many cases the recombining plants were placed in local ownership with New Zealand contracting to supply the raw material and to provide technical expertise. Markets for spray-dried milk powder spread during the 1960s and 1970s to Japan, Arab states and South and Central America.

The drive for a wider range of consumer products from milk forced up the tempo of amalgamations of co-operative companies in New Zealand. Farmers accepted that the capital demands for modern machinery rendered small dairy factories run by home-town or even provincial companies impossible in the context of world trade with a shifting demand for various milk products. Small co-operatives with ageing plants could no longer offer their member-farmers a payout that would match the amounts available to those belonging to expanding companies investing in large, efficient, flexible plants.

In some cases resistance to amalgamation arose among communities which had nostalgic loyalty to their local co-operative with allegiances going back several generations. The original factories had been spaced according to how far a horse and cart could carry milk over not very good roads once a day. Some of these small companies wanted to stay separate long after it made sense to do so. If they tried to increase flexibility of production in their own small area, they tended to overcapitalise and fell behind in payout levels anyway. Gradually, economic reality always prevailed. The basic commodity remained cows' milk but, increasingly, manufacturing capacity needed to be large and flexible to shift its production of milk's derivatives – butter, myriad cheeses, milk powders, caseins and a range of other specialist products – in response to demand among dozens of markets.

Since the 1960s roads had been upgraded throughout the country, enabling the big tankers to pick up milk on spider-web collection runs. In 1961 dairy factories numbered more than 165 but by 1973 had dropped to seventy-five. In 1971 Taranaki was home to forty companies. Twenty years later there was one.

The next big challenge, thirteen years after Britain joined the EEC, was the deregulation of the New Zealand economy. The comforting rug of subsidies with its intricate pattern of payments was pulled out from under farming over two seasons. During that period, the Dairy Board legislation was amended nine times as farming politicians scurried back and forth to the Beehive. Farmers with high debt against their properties were soon in crisis and some foundered. Dairy farmers had advantages, though, over meat and wool growers: the industry account that supported prices was in credit; most dairy farmers and their families were milking cows themselves with herd sizes then around 150; they were likely to have less exposure to debt because of the traditional stepping-stone approach to farm purchase, via share-milking; and they had better cash flow.

After the Northland Co-operative Dairy Company amalgamated with Kiwi, the industry had two dominant companies left in the late 1990s, each based in one of the great national dairying regions: Kiwi in Taranaki and the New Zealand Dairy Group (NZDG) in the Waikato and Bay of Plenty. Kiwi and the NZDG then merged in 2001 to become Fonterra. Only a handful of small companies stayed outside the giant conglomerates, Westland Milk Products and Tatua Co-op Dairy Company, and some set up soon afterwards. Their aim was to supply niche markets mainly for the domestic market. Otherwise, the industry's manufacturing capacity was taken within the Fonterra monolith to achieve economies of scale and prevent the likelihood that two companies would compete for price in export markets and diminish the value of both.

The Kauri dairy factory, north of Whangarei, was built by the Northland Co-operative Dairy Company, then absorbed by the Kiwi group, and subsequently Kiwi joined with the New Zealand Dairy Group to form Fonterra.

Following pages: A disused dairy factory in Taranaki.

Dairy farm entrepreneurs

Colin Armer's father and mother had their own dairy farm in the Bay of Plenty, and in the late 1970s he began share-milking for them, using his first car as security. He stayed on for the new owners when the family farm was sold. He then took the traditional route to owning a farm of his own: borrowed money, built up his herd numbers, increased his cash flow. Then in 1982 he sold a portion of his herd for a deposit on his first farm – a forty-hectare property, with which he won a regional dairy farm of the year award for demonstrating productivity and profitability as a business.

He reckoned the shock of the Labour Government's restructuring, with its consequent high interest rates, was the trigger that led to him and wife Dale taking the sort of smart chances that within fifteen years put them in the extraordinary position of being one of dairy co-operative Fonterra's largest shareholders. By 2005 they and their companies controlled eighty-six dairy farms with 70,000 cows in both islands and across the Tasman, and they were growing the herd by ten per cent a year. Share-milkers owned about 20,000 of the cows in milk and the companies owned 50,000. The operation provided about twenty-one million kilograms of milk solids a year. The national dairy landscape was changing dramatically. The Armers were not alone, with family corporates and other equity groups growing at a similar rate.

Back in the 1980s the Armers leased a lot of land and reared a lot of stock, and when an upturn in the market occurred in 1989 they bought a second farm and then others until they had enough seed money to move into converting sheep farms to dairying. They bought 300 hectares of land near Te Puke which had just adopted a drainage plan, and spent time developing it – 'Stretching our banker's imagination again.' In simple terms, Colin said, it worked like this: 'You buy a sheep farm for a million dollars and spend another million on it, and then you wake up at the end of the development period, and it's worth three million dollars, so you do another one.'

The Armers did some things differently. For example, when everyone was breeding bull calves from cows deemed surplus to their milking herd, they reared dairy heifers, and when dairying picked up in the 1990s they had all these cows to sell. That meant a deposit on another farm – 'And off we went.'

In 2000 the Armers and three others formed Dairy Holdings Ltd, bought a majority shareholding in the Tasman Agricultural Co with interests in Australia and turned it into Tasman Farms, now a subsidiary of Dairy Holdings, a private company. Once they had expanded to the degree that they had to employ large numbers of staff they had to learn new management skills, gradually setting in place a structure that divorced them from day-to-day concerns. Dale Armer took care of office administration, accounts and budgets. Colin ran four farms with four managers on them, two supervisors looked after another ten farms, and in the South Island he communicated with a general manager under whom worked seven supervisors across the South Island and in Australia. Salaried managers had the career

Dale and Colin Armer knew that once they had expanded and had to employ large numbers of staff, they had to learn new management skills.

opportunity of moving into a share-milking arrangement.

In the early 1980s Colin had visited the research station at Ruakura and found his farming practice was similar to the grazing management pasture-based system developed on No. 2 Dairy. For the following ten years he took busloads of staff three or four times a year to study the work being done there. No. 2 Dairy was based on twenty-cow farmlets and the Armers adapted the research to thousand-cow farms. 'We have varied the systems a little bit for climatic conditions and scale and made some adjustments, but primarily the observations and training I got at Ruakura through the 1980s was driving the business through the 1990s,' Colin said later. Some farmers adopted systems he considered unnecessarily complicated, and risked wasting pasture: 'We don't make much silage. We concentrate on quality pasture and maximising the profit from the pasture a farm can grow . . . Our business was allowed to grow because we kept it simple and repeatable.'

As always, Colin Armer is still thinking ahead in the new century. While most others were building bigger and bigger herds, he was moving slightly in the other direction. He said all the benchmarking they did on financial returns showed they got better results from an 800-cow farm than from the bigger herds, mostly because the management of such a herd was more effective; so they moved towards that size as the optimum.

The Jersey cow prevailed for decades until overtaken by the Holstein-Friesian in the 1980s.

Before that final merger, the concept of a duopoly with Kiwi and New Zealand Dairy Group competing had been held by some to be the ideal structure, one that would foster competition in both the domestic and overseas markets and force the efficiencies that competition brings. The Commerce Commission's first ruling was against a merger but the government supported the monopoly and approval was finally given on condition that the new, big company took some sell-off steps to ensure domestic competition.

Thus Fonterra was born, a huge co-operative company, one of the top ten dairy companies in the world and the biggest exporter (to 140 countries), responsible for one-third of the world's dairy trade. It gave itself a dreamt-up, neutral name that bespoke the company's determination to hear no parochial echoes – Fonterra has no linguistic or semantic attachment to any language. Also changed were the rules of engagement for farmer-members. Previously, they entered their co-operatives with minimal fees and were paid out on the volume of milk sold; but variable joining fees were now imposed based partly on the size of the supplier.

As the co-operative dairy companies were getting bigger so were farms and the herds they carried, and the composition of those herds was changing too. The most successful breed in the beginning was the milking shorthorn. The first Ayrshire bull arrived with the Otago settlers aboard the *Philip Laing* in 1848. 'Rob Roy', as he was called, had an immediate impact on dairying in that province and Ayrshires have been a positive influence in the national herd ever since. Another early importation was the Dexter, also known as the Kerry, but it disappeared. Two breeds have dominated the industry – Jerseys and Holstein-Friesians.

The first Jerseys arrived in 1862, brought in from Jersey by Thomas Syers of Wanganui. Other importations followed, and it was in Taranaki that the merit of the small-framed, fawn-coloured cow with its rich lode of butterfat was first recognised. Herd testing in the Waikato underlined the extraordinary productivity of the animal. By the middle of last century Jerseys numbered 1.5 million, and Friesians about 110,000, slightly more than milking shorthorns. Jerseys received the imprimatur of government researchers immediately after World War Two after prolonged testing. But this was a time when payout was dominated by butterfat

The milking ritual

Once a dairy herd size exceeded, say, fifteen or twenty cows milking machines became economic and a cowshed had to become more than just a shelter. A walk-through shed with four, six or more bails became essential as soon as a farmer could acquire the capital. The cow's head was held in a frame, her outside foot tied to a rail with a leg-rope across the inside foot, so the animal couldn't easily kick. The milker cleaned the teats, applied the cups and, when the machines had finished, 'stripping' by hand got the last of the milk. The ankle cord was then untied, the head frame opened by means of a long pole and the cow walked along a race to a holding paddock. The gate to the yard was opened and another cow brought in to the empty bail. The walk-through provided an efficient milking system for small herds and a few were still in operation late in the twentieth century.

The next development, after World War Two, was the herringbone shed with cows entering in single file, mostly slightly angle-parked, facing outwards, on each side of a central pit from which milkers applied the cups, working at shoulder level to speed up the process and avoid the constant bending associated with the walk-through. The first herringbones were based on designs imported from overseas, which were then adapted to New Zealand conditions. Hand 'stripping' was abandoned so the cows flowed in a fairly seamless, circular motion from the yard, through open stalls and out again. Herringbones – with their greatly improved time-and-motion for milkers and throughput for cows – spread rapidly as herd sizes grew. Herds of 500 cows could be milked efficiently by two people.

The next big improvement to meet the needs of ever-growing herds was the rotary system. As the milking platform revolves, driven by an electric motor, each cow moves into a bail and the cups are placed on. After one, timed, revolution (from eight to ten minutes, depending on the time of the season) the cups are removed (in some cases automatically), the cow backs out, leaving the bail vacant for the next cow to step in as the platform edges around. The rotaries vary in size and style but the basic system is similar and cow throughput jumps impressively for big herds.

The development of the rotary turnstile shed was a tribute to the innovative thinking and confident stockmanship of Taranaki dairy farmer Merv Hicks. All earlier milking sheds had been designed around the principle of a cow walking straight into a bail and then moving forward again once the milking was completed. Devising a rotary system in which cows could continuously progress in this way posed what most saw as insurmountable technical difficulties. Hicks wrestled with the problem as he considered a new shed for his 100 cows after his old shed had been condemned by a dairy inspector in 1967. He had two years to build a new one.

A simple solution to the rotary system would be to have the cows back out of their moving bails after their turn, but this flew in the face of what had become conventional wisdom: it would be cumbersome, and unnatural for the cows. However, Hicks could not see why they couldn't be taught to modify their behaviour; so he decided to teach a couple of cows he milked through the winter to back out from their bails in his walk-through – it took only days. He had an engineer friend draw up plans, got tentative approval from the Dairy Board, and the first turntable shed began operating in 1969. The cows learnt to back off in a day, Hicks said later. What surprised everyone was that the cows behaved better than in herringbone sheds. Hicks said that 'even the ornery animals' weren't making trouble because once in the separate, moving bail they enjoyed their own personal space. That cool confidence in his appreciation of cow psychology encouraged him to take a $10,000 gamble on a new rotary design.

Since then more than a thousand New Zealand dairy farmers have installed rotary sheds and the system has been exported to the major dairying countries.

The herringbone (left) succeeded the walk-though cowshed successfully for decades but has since been superseded by the rotary which milks a cow in one cycle.

content for the making of export butter. As the trend changed to milk volume with other ingredients of milk at least as important, the predominant breed became Friesian.

In 1972 Jerseys still comprised fifty-eight per cent of the national dairy herd of about 2,900,000 cattle, Friesians thirty-four per cent, and Ayrshires three per cent. By the end of that decade Friesians outnumbered Jerseys, and the two breeds have staved off challengers from other dairying breeds brought in since the 1980s. The big-boned, black-and-white Friesians gave easily the highest volumes. Friesians were also especially valuable for crossbreeding with beef sires for dairy beef. Gradually they gained the ascendancy. The prevailing view now is that a place exists in the national herd for any breed or cross that is most efficient for a particular productive purpose; but now that butterfat is not the main requirement, the big Friesians provide milk volume rich in solids not fat.

The industry in New Zealand has depended almost entirely on four cattle breeds: shorthorn, Ayrshire, Jersey and Holstein-Friesian – all of them, except the shorthorn, eponymously named for the geographical regions in which they were developed.

The earliest herds were dominated by the shorthorn, which for a long time was the most important cattle breed in New Zealand – a good milker which also produced excellent steers both for draught work and for meat. Milking shorthorns populated many dairy herds around the world and a number of successful beef breeds have been fixed from shorthorn crosses. The breed was developed in England's Durham County from a cross with imported Dutch cattle, and then one breeder selected for milking ability and another for beef. Shorthorns were still called Durhams when they arrived in New Zealand in 1814 with missionary Samuel Marsden as a gift from New South Wales Governor Lachlan Macquarie. They spread rapidly through the country, serving as multi-purpose animals. As in England, some were selected for beef herds and others for their capacity as milkers. By the 1920s more than half the dairy cattle in New Zealand were milking shorthorns, but later that decade they began to give way to the Jersey. By the beginning of the twenty-first century, as the bicentenary of the breed's arrival here approached, so few were left in dairy herds a move was afoot to

Milking time on the Bennett's farm, Ngatea.

preserve some animals in recognition of the breed's long-standing service to New Zealand farming.

The Ayrshire was developed during the eighteenth century when the rudiments of crossing cattle and selecting for certain traits were becoming understood. Before the breed was set in Scotland's County of Ayr, crosses involved some European strains – perhaps some from the Channel Islands, and certainly Teeswater cattle from Britain's Tees River Valley, which were also progenitors of the shorthorn. On the way to its final fixing, the breed had been called the Cunningham and the Dunlop. The conformation of the Ayrshire, and especially its udder, is admired, and it took its place in New Zealand dairy herds as an efficient converter of grass to milk.

The Jersey, with its feminine face, lovely moony eyes and fawn colouring, came originally from the small Channel Island off the coast of France, although the island was once known as Alderney as was its most famous cattle. Some claim the animal has been a pure-bred for six centuries, and it was certainly known in England as a superior producer of butterfat about 250 years ago. During the nineteenth century Jerseys were imported by every country with a dairy industry and proved adaptive to a wide range of conditions. In America it was bred up in size and lost some of its delicate conformation – with straight top lines right through to the rump and deep in the body. It is an extraordinarily productive cow for its size and it reigned as the queen of the New Zealand dairy industry as long as farmers were paid solely on the basis of butterfat content. At the end of the 1950s, seventy-five per cent of New Zealand dairy cows were Jerseys. An interesting contrast exists between the female and the male, which, although smaller than most dairy bulls, is tough and aggressive.

The Holstein-Friesian has supplanted the Jersey but the Jersey remains a major component of the national dairy herd of approaching four million, both as a pure-bred (about 15 per cent) and as a Friesian-Jersey cross (about 25 per cent). Another Channel Islands breed, the Guernsey (also eponymous), has been present in New Zealand for some time but has yet to make an impact.

Friesians originally came from the West Friesland region of Holland and Holsteins from Holstein, Germany, 300 kilometres to the north. However, the breeds are now indistinguishable as many countries have used animals from both regions to develop the same black-and-white breed. The first Dutch Friesian cattle were imported by famous Canterbury farmer John Grigg in 1884, and this is the reason New Zealand farmers have tended to use the name 'Friesian', even though the breed name here is officially New Zealand Holstein-Friesian.

Since the 1960s North American Holsteins have been introduced, at first through Canada, and more lately from the United States. The North American black-and-whites – farmed mostly indoors and not under the intensive grazing system used in New Zealand – are huge animals which produce impressive volumes of milk. This introduction made itself felt in New Zealand production over the last three decades of the twentieth century and saw the Holstein-Friesian become the predominant dairy animal. However, in the United States, the breed had an average lactation of only around two years because animals had to be culled young when they did not get pregnant. This fertility problem was generally passed on to herds here and resulted in a decline in the national average lactation. Thus the industry turned its attention to breeding for fertility in a bid to enhance the already established economic value of the Friesian.

Another European dairy breed that is well established in many countries and known for its long lactation life is the brown Swiss, introduced in the mid-1970s. Another Swiss breed, the Simmental, was introduced into New Zealand as a beef animal in 1970, but was considered a dairy animal as well in Switzerland. At least one Taranaki farmer began milking Simmentals, growing his bull calves as beef.

New Zealand farmers have shown a starkly pragmatic approach to animal breeding, helped by the brilliant recording systems of Livestock Improvement and also by milking-shed technology that increasingly enabled them to monitor the production of each cow at milking time. Thus they have bred and culled according to the productive capacity of their individual animals and not according to its breed or superficial conformation. Friesians came into favour entirely because of their ability to produce high volumes of milk. The number of Friesian-Jersey crosses steadily increased because of individual productive capacity.

Keeping it simple

Geoff and Marion Mathis put together what he calls 'a jigsaw' of eight dairy farms over thirty years from 1970, and by 2005 they and their share-milking son carried 1100 cows on the 349-hectare property, milked in a rotary shed built on the centre farm – 'the crown jewel, in the hub of the wheel', as he put it. That way the cows didn't have to walk too far and therefore didn't suffer the feet problems of some of the bigger herds. The cows were all Jersey, Friesian or a mix of the two; and the operation was largely self-sufficient, supported by about thirty hectares of grain for supplementary feed.

Mathis's grandfather arrived in New Zealand from Switzerland in 1913, the eldest of eight children, and a year later he married a woman from East Germany, also one of eight children in a family that had come here to escape the stresses of the impending First World War. The Mathis family farmed in the South Waikato region from then on. Geoff's father bought the first 133 acres of the present farm in 1952. Geoff married Marion in 1970 and they bought an adjoining seventy-four-acre farm with a hundred per cent borrowed finance, and share-milked on his father's farm, which he later bought. Thus began the aggregation that reached its peak and was worth around $13 million in 2005. At that time, Geoff and Marion still lived on the property but had decided to pull back a bit. The Swiss connection remains. Cousins are still on the family farm there, milking twenty-eight cows.

I suggested that twenty or thirty years ago, you could run a dairy farm with a cash book and a herd book but the range of skills demanded now is much greater. 'No, I don't think so,' Geoff said. 'Farming is simple. Individuals make farming complicated. I could work from a recipe book McMeekan used in 1952. It hasn't changed very much . . . You've got to look at farming and say, "Where do you want to go and what do you want to do?" My success in farming has been based on maximising capital gain. My cash flow has never been extremely high . . . I've grown through borrowings and pushing boundaries. Others have done it my way and others have done it by concentrating deeply on maximising performance per cow per hectare. I've made sure I had good averages in everything – good averages in every performance, not sitting in the top ten perhaps but sitting in the top fifteen per cent. I'm satisfied I've taken advantage of all my skills.'

Mathis maintained he always had a plan in his head. 'I didn't need to sit down and write my goals. In 1984 we purchased a sixty-acre run-off. Then Muldoon called the snap election, the world changed overnight and we finished up having to finance at twenty-six percent for three years. We didn't put any fertiliser on. I think we got through two and a half years, paid a penalty and got out of it. We had bought time.'

Geoff and Marion Mathis on their property near Tirau which they put together as 'a jigsaw of eight dairy farms over thirty years from 1970'.

The Mathis herd of eleven hundred cows walks to the rotary shed built on the centre farm, 'the crown jewel, the hub of the wheel'.

Above: Breeding bulls in Taranaki in 1911, a Holstein-Friesian (left) and an Ayrshire, historically two of the main components of the New Zealand dairy herd. The Friesian was originally prominent in town supply herds because of the volume of milk it produced.

Opposite: Farmers have always taken full advantage of mechanisation as does Lorcan Kirwan on the Bennett's property near Ngatea.

Meanwhile, early research into desirable traits for once-a-day milking hinted that the Jersey might make something of a comeback because under those conditions it initially did very well.

But it was herd size that changed most in the modern era. The days when a family could make a good living from thirty or forty cows disappeared. Technology paved the way. Milking machines became increasingly efficient. Hand 'stripping' of cows ceased and herringbone sheds proliferated and increased throughput. The pits enabled milkers to avoid back-breaking bending to put on and take off teat cups. The national herd of dairy cows grew from 370,000 in 1900 to 1,850,000 in 1950 but the average herd size increased only from twenty-plus to thirty-something during that half-century. After that, gradually at first and then very rapidly indeed, the demographics of dairying – or 'bovigraphics' to coin a word – changed as the number of herds declined, growing individually ever larger, and dairy farms began to spread into country traditionally avoided because it lacked adequate rainfall spread widely through the year.

In the thirty years from 1975 the number of herds dropped from more than 18,000 to 12,500; average herds grew threefold from 100 cows to more than 300; a thousand cows became common and the biggest herds were two or three times larger than that. The explosion in growth occurred in the last decade of that period with total cows surging from 2.8 million in 1995 to 3.9 million in 2005 and average herd size from 193 to 315. Average milk solids per cow graphed up as well, from 271 kilograms to 308, with some small annual dips caused by unseasonal weather.

Again, technology gave impetus to the trend. The rotary shed, developed by a New Zealand farmer, meant more cows could be milked at a time with less labour, although labour remained the problem it had been for many years. But new urgencies were emerging that would require scientific attention – among them: foot soreness in large herds on big farms having to walk long distances to the shed, and losing some condition in the process; declining fertility; and environmental pollution from the intensity of modern dairy-farming practice. Between 1995 and 2005, the average number of cows per hectare grew from around 2.5 to 2.8. That

Tradable shares or co-operatives?

Industrial, rural and retail co-operatives were fashionable in the early years of the twentieth century when the dairy industry was developing, and not only in New Zealand but in other Western countries. The concept was ideally suited to dairy manufacturing and worked well for farmers right through the century.

However, demutualisation with the issue of tradable shares was raised on occasions as a possible change in structure. In many cases advocates were ideologues who had philosophical problems with what they saw as a form of syndicalism – except that the plant was not owned by workers but suppliers. Such a move was also raised, though, by some more pragmatic farmers who realised that, as the cost of land escalated, an increasing number of farms were owned not by nuclear families or family corporates but by groups of people, many of them investors. About 11,000 farms supplied Fonterra five years after it was formed, down from 13,000 as herds became bigger; but as shareholders, or shareholding entities, numbered around 10,000, multiple farm ownership was obvious. Also, it was known that some farmers, or investors, owned more than one of those entities. Although Fonterra said it had no way of knowing the extent of multiple farm ownership, one estimate by an industry leader was that a third of the supplying farms were owned by about a hundred farmers; a trend, he said, that was increasing.

The question sometimes asked was whether investors would need bigger returns than they could pull from the company payout. There was also the worry that as the company expanded globally local suppliers' interest would be submerged by expansion, especially if other countries could provide cheaper milk in larger volumes than New Zealand. Some observers felt that if the industry became an expanding international consumer business it might need to pull in equity from outside.

But the prevailing feeling in the industry was that the long, strong co-operative tradition had served New Zealand well, and the question of demutualisation was quietened, at least for the time being, by Fonterra's capital appreciation in the early years of its formation. Also, most farmers worried that a concomitant of demutualisation would be a loss of market power for them, and that short-term financial advantage would lead to lower returns in the long run.

meant more nitrogen entering into waterways from fertiliser run-off and from cattle urine and manure.

Dairy farmers were unhappily resigned to incurring costs in the fight to keep waterways free of pollution as yet another change in farming practice loomed: milking once a day instead of twice.

Many farms had moved to once-a-day (OAD) and more were investigating a change. The advantages were the obvious lifestyle improvement for farmers and families, lower labour costs, and less strain on cows getting to and from the shed. Early research by the user-owned Livestock Improvement co-operative demonstrated that some cows adjusted to OAD better than others and the company developed a breeding index ranking bulls and cows on their performance under OAD conditions. Farmers making the change needed to patiently modify the composition of their herds to get the best results. Milk production losses in trials were variable – anywhere between six and eighteen per cent. In some cases farmers compensated for this variation by more intensive stocking to get better use of the grass. Overall, the advantages of OAD milking were strong enough to drive a firm trend.

However, the national average of lactations for cows was dropping as the new century dawned, with the percentage of cows coming into calf declining. The reasons were not entirely clear and it was a problem afflicting herds worldwide, perhaps because of the rapid international transfer of genetic material. The situation demonstrated that the focus on genetics for advances in milk volume and quality could result in a distraction from other desirable traits. But the New Zealand industry is so well organised now it can shift focus with relative speed towards breeding for higher fecundity. One approach taken by farmers is cross-breeding for hybrid vigour by using top-quality Jersey semen over their Friesians, thus enhancing in-calf percentages immediately.

In the days when a herd of forty or thirty or even only twenty was an economic unit, milking in the early morning and late afternoon was an unremitting family chore, but before milk yield was precisely measured, the family knew the generous milkers from the parsimonious, the irritable cow from the tractable. The good farmer built up a family profile in his head, perhaps, for a small herd, or in a dog-eared notebook that served as a herd book. He culled not only the cows which did not come into calf, but noted those that took more than one service to become pregnant; although even after herd testing and artificial insemination became available he would keep a bull, if he could afford it, for a second chance or for the cows that came in heat late, or he might share a bull with a neighbour.

In the very big herds now, the cow is a number, a stock unit, where once she was also a name, a personality with a private reputation. The farmer at first built up a herd from affordable local cows and bulls. Compared with the modern cow grazed rotationally, their performance was poor. But the government- and industry-sponsored artificial breeding scheme exponentially lifted performance.

Dairying also has a long history of using science to its profit in New Zealand, and although the level of research – especially what is called blue-sky research – declined from 1986, advisory organisations such as Livestock Improvement and Dexcel doggedly tackled practical problems and worked towards the future. The moves – the dreams, if you like – that the industry was pursuing included robots to do the milking, perhaps in mobile milking sheds that would go to the cows instead of the cows having to go to the shed. Robots were already in action in the closed-in sheds of Europe.

During the boom decade from 1995, good prices convinced many farmers to switch from fat lambs and crops. In some cases the land was marginal and some observers worried for the future. The eastern coastal strips of both islands had never been favoured for dairying because they were generally drier than the western and central regions. But as the new century opened, some of the country's largest dairy farms spread over Canterbury, where the available water for irrigation was being contested by other groups including power generation companies and recreational fishers and boaters.

Nationally, in the rush to the future, perhaps the most serious concerns were water quality and availability, and the industry's interface with the environment which would require new techniques to overcome pollution – for example, new ways to fix nitrogen in pasture. Farmers were aware that handling these issues would result in increased costs.

Opposite: The Ayrshire.

Above: The Friesian.

Left: The Jersey, with its fawn colour and moony eyes, is the most tractable and liked. It was favoured in New Zealand for decades because of the prodigious amount of butterfat it produced from within its small frame.

CHAPTER FIVE

THE RE-EMERGENCE OF LAMB

Opposite top: A traction engine ready to pull away with a load of wool bales in the days when wool helped keep the New Zealand economy afloat. A shearing gang poses for a photograph with the fruits of their labour.

Opposite bottom: Wool buyers inspect fleeces available for sale before the bidding starts. Wool was a truly international commodity and overseas buyers had representatives in New Zealand, but erratic prices and uneven quality gave synthetics an advantage after World War Two – an advantage they have never lost.

IN THE OPENING decades of New Zealand settlement by Europeans, wool grown on sheep grazing native tussock (and, later, English grasses) in the country's big paddocks was as valuable as the gold thread spun from straw by Rumpelstiltskin's slave. It was an international commodity, and even if its price came and went over the years it stayed high enough for fortunes to be made from it, mainly in Canterbury, Marlborough, Hawke's Bay, the East Coast and Otago.

That state of affairs changed after World War Two when synthetic fibres became more plentiful, cheaper and available for variable use. Synthetics had the market advantage of being able to meet the required specifications at prearranged prices, whereas wool was sometimes not consistently true to label when it arrived at the manufacturing plant, and prices fluctuated from week to week. The wool-producing nations combined, with some success, to market wool under the imprimatur of the Woolmark, which tried to guarantee even quality of product. Fine wool held up better than the coarse wools which were predominantly what New Zealand grew for the burgeoning carpet industry.

When Britain was subsumed by Europe, wool did not suffer the same sudden, disastrous consequence as meat and dairy products. But, as the competition from synthetics spread, the price settled at a level that was closer to the value of Rumpelstiltskin's straw than the gold spun from it.

By the 1960s New Zealand sheep flocks included close to thirty breeds, among them animals that had come in originally from Britain, some that had

Wool buyers were highly paid men who bought on order. They travelled the country attending auctions in the wool centres and brought sheep farmers joy or grief in good times or bad. The whole system has now changed.

been developed here, and others imported through Australia. The merino had just held its own on the dry hills and high country of the South Island, and others such as the Cheviot, Corriedale, halfbred, English Leicester, Border Leicester and Southdown were still present in numbers. But easily the most ubiquitous animal was the Romney, comprising near enough to three-quarters of the national flock, a successor early in the century to the hardy Lincoln. Imported as the Romney Marsh from the 1850s, the local version developed into a type of its own and became the standard dual-purpose sheep with a good carcass and sound, strong wool. Other Romney offshoots were making their mark during the 1960s – Perendale (Cheviot/Romney) on the hard country and Coopworth (Border Leicester/Romney) on the foothills and flats. The Drysdale was also bred from the Romney – which carried a gene for hairy fleece and horns – primarily to provide exceptionally coarse wool as a component of carpets. It was a niche that had previously been filled by Scottish blackface not available in this country. The Southdown arrived in New Zealand early and was the most common terminal sire,

The battle of the breeds

Pastoral animals were originally brought to New Zealand from Britain, often via Australia. The first sheep to arrive in any numbers was the merino. Its fine wool was so prized that centuries ago smuggling or attempting to smuggle the breed out of its native Spain was a capital offence. But strong-wool breeds, notably the Romney Marsh, became more important to New Zealand farmers and, like others, were controlled by breed societies that looked back to their origins in England to ensure the breeds remained vigorous, 'pure' and true to type.

The system worked in the following way. Stud breeders registered their ewes and rams with the appropriate breed society or association providing officially recorded pedigrees. In front of each flock book was a long and detailed description of what the ideal animal of that particular breed should look like – its 'conformation'. These breeders either sold rams direct to commercial farmers or on to middlemen, if you like, who multiplied the animals and then sold them on to commercial farmers. Commercial farmers were not allowed to sell rams for breeding.

This mystique of conformity and purity was reinforced by the purchase from time to time by stud breeders of expensive, prestigious sires from the breed societies in Britain to renew the 'pure' genes; and British stud breeders were often brought to New Zealand to judge local stock at A & P shows to make sure breed standards were true and consistent. The system was such a brilliant exercise in marketing by the British breed societies that it became international, especially among those countries within the empire. Thus the United Kingdom became the lucrative stud farm for the world.

The breed concept even spread to humans. The immigrant population in New Zealand was even more narrowly homogeneous than the livestock. Almost all the early settlers came from England, Scotland or Northern Ireland, and they brought with them a belief in the moral power of British Protestant righteousness that was all the better, they felt, for regular cultural infusions from the Mother Country. As late as the 1930s, Ewen Alison, businessman, former Mayor of Devonport and then Takapuna, MP and member of the Legislative Council, could write:

New Zealand has a heavy atmosphere. It is about fifteen pounds to the square inch. This heaviness of atmosphere tends to make people sluggish and slow moving. This applied more particularly to Auckland than it does to Wellington and the South Island . . .

Our young people of the second, third and fourth generation are not so virile and robust as their parents and grandparents. We are a retrograde nation physically, and it is necessary to have infusions of blood from the Homeland to keep up our excellence. Our horses, our cattle, our sheep and our dogs require the same restoration . . .

While very few people truly accepted the degradation of human colonial genes, the concept of renewing pure British livestock strains persisted into the second half of the twentieth century. However, some practical farmers gradually became aware that the economic performance of their animals was not always consistent with their so-called purity. Ruakura scientist Dr Percy McMeekan and his associates enraged breed societies by telling farmers that they should breed their Jersey cows basically for butterfat production, not for their comely conformity. Herd testing of dairy cows was, by then, well under way and the evidence was compelling.

A number of farmer-breeders were at the forefront of the campaign to use scientific methods to increase milk production, notably Jack Ranstead, the founder and first president of the New Zealand Society of Animal Production, and a neighbour, H. E. Annett, bred up a superb herd of pedigree milking shorthorns, and built what was described as the best private library in the country on genetics and animal production. His house was a magnet for scientists from around the country and he had an extraordinary influence on animal breeding and production, and not only of dairy cattle.

Then, in the 1960s, Professor A. L. Rae of Massey Agricultural College helped enlightened farmers to set up the National Flock Recording Scheme (NFRS) which offered to stud breeders, and anyone else interested, the chance to record for the first time in an organised way which brought early maturity to lambs and allowed them to lay down fat for the sort of blocky carcass beloved at the time by the British market.

The industry was set in its ways, with Romney breeding ewes shorn for their wool and providing lambs for the meat trade. Trouble was, the Romney had changed shape from the original imports and was having fertility problems, and lambing percentages were usually a shade under a hundred. Then demand for fat lambs waned in favour of lean meat. One industry executive says that in the early 1970s, when he was working in Britain, the carcass that won the Best New Zealand Lamb competition at Smithfield was a Southdown cross which had a layer of fat all over it. 'Prime', with a covering of fat, was the grade that was tops in what was called the fat lamb trade. Another grade, called 'seconds' did not have as much fat and, as tastes changed, the seconds began to fetch the higher price.

About the time this was happening – as a consequence of falling meat prices in the wake of the EEC embracing Britain and few alternative markets opening up – successive New Zealand governments

performance data such as lambing percentages, live weights and fleece weights. Coopworth (Border Leicester/Romney) and Perendales (Cheviot/Romney), new breeds fixed during the 1950s, were taking their place in many commercial flocks and the Coopworth Society made it mandatory to be on the NFRS. But the impetus came mainly from commercial farmers who tagged and recorded the performance of their sheep for processing by NFRS to identify the top ewe lambs for future replacements. One of the main considerations for the sheep farmer was to identify ewes with good lambing percentages and easy births to fit in with the easy-care New Zealand farming style.

The stud breeders' income was hit as farmers were surprised to find how high was the quality of their top ewes and rams and began breeding up within their own flocks, and even selling rams to other farmers who wanted sheep bred for productive lives in the economic reality of the farm rather than in small studs.

The next move was the Group Breeding Scheme (GBS), in which some groups of commercial breeders identified their top ewes (usually two-tooths) and contributed them to a central flock, usually of a group member's farm, for ease of recording. Rams bred in the nucleus then went back to be used on contributors' farms at a ratio of one ram for four contributed two-tooths. The nucleus farmer was paid a management fee for the extra work of recording.

As this revolution developed, stud breeders continued to miss out on sales, breed associations lost registration fees, and stock and station agencies got no commission. Lobbyists sought to frustrate the new developments and to curb the support from government-paid scientists, but breeding for economic traits became an unstoppable trend and was essential to survival once farming subsidies were eliminated in the 1980s. The whole livestock industry was transformed. In the end, so were the breed societies, which nowadays provide full performance recording services and advice on all aspects of breeding.

Massey Agricultural College Professor A. L. Rae set up the National Flock Recording Scheme which surprised farmers with the quality of their top ewes and rams.

Above left: A stud merino ram. Merinos' numbers declined as the twentieth century progressed but they were important economically in the high country. Their fine wool, valued in the garment industry, commanded high prices.

Above: Texel/Perendale crosses. Big and fertile European sheep like the Texel, imported since the 1970s, have earned a valued place in the national flock, increasing carcass size and enhancing the lambing percentage.

began increasing subsidies to sheep farmers, especially via Robert Muldoon's 'ewe retention scheme', to keep the industry going through what they perceived to be a cyclical downturn and an overpriced currency. The effect was an alarming increase in the national flock, from forty-seven million in 1960 to sixty million in 1970, and more than seventy million in the early 1980s. Then, as the supports were withdrawn in the mid-1980s, sheep numbers plummeted – to fifty-eight million in 1990, forty-two million in 2000, and down to thirty-nine million in 2005. What was extraordinary – and a tribute to farmers' acumen, foresight and efficiency – was that those thirty-nine million sheep produced more export lamb, about 358,000 tonnes, than in 1987 when the total from more than sixty million lambs was just under 350,000 tonnes. The reasons were a soaring lambing percentage – up from an historical figure of around one hundred or slightly less in 1990 to more than 120 fifteen years later. All because of the introduction of new European breeds.

A downside of the improvement was that sheep got so big so quickly shearers found them hard to handle, and investment in technology had not kept up for the industry to work towards a solution. This slowed the process and imposed heavy demands on the ACC and consequently on premiums. The problem was compounded by the gradual ageing of shearers, with a declining number taking up such a tough job.

When the National Government's supplementary minimum payments for sheep (SMPs) were taken away in the mid-1980s, farmers started looking at efficiencies and costs – and realised that the cost of taking the wool off a sheep was economically becoming less attractive than ever. Most of them figured lamb was going to provide a better base return for the costs involved, so the emphasis of their farming practice was directed towards getting more and bigger lambs from the declining number of ewes, with wool becoming virtually a by-product. The following fifteen years confirmed the decision as the right one for the time. It also gave farmers confidence that they could face world markets.

Interestingly enough, Ruakura scientist Dr A. H. Kirton had said in 1964 that New Zealand would have to increase its pastoral animal population by four per cent a year just to keep up, and noted that the application of scientific principles of breeding in sheep and beef cattle was forty years behind the dairy industry. In fact, the economic productivity of the pastoral sector – not animal numbers – increased four per cent a year over the twenty years from 1985; and once they shucked off lazy practices, the sheep and beef industries were contributors to this performance. What gave lamb its productive thrust was the introduction of exotic breeds to boost flock fecundity and increase carcass weight.

The lift in lambing percentages followed the importation of first the Finnish landrace and then the Texel, both of which also added to the body size of progeny when used as terminal sires over traditional local breeds, including Romney. Romney still made up half the national flock in the new century. Texel and Finns both came from northern Europe.

Opposite: Walter Peak High Country Farm on the western shore of Lake Wakatipu, opposite Queenstown, is both a sheep station and a tourist attraction. Visitors ferried across the lake by the TSS *Earnslaw* may watch sheepdogs and wool spinning in action, as well as shearing, and have farming practices explained.

The flamboyant Godfrey Bowen is introduced to the Queen at Napier during the 1954 Royal Tour, after a demonstration of his and his brother Ivan's shearing technique and speed. The Bowen brothers (Ivan in the background) turned shearing into entertainment for townies all over the world.

Shearing as theatre

The Bowen Brothers – Ivan and the more flamboyant Godfrey – were instrumental in bringing about a revolution in shearing technique and transforming the hard graft of rural work into a sport and entertainment for townies all over the world. The brothers first took black singlets into the limelight in the early 1950s with world shearing records. Godfrey, short and thickset, ideally built for the job, had refined his shearing technique since the age of sixteen and also possessed the kind of charisma that made him a star. He elevated the job of shearing first into a dramatic competitive sport, and then into an international entertainment.

The brothers were two of four sons of a Hawke's Bay sheep farmer, and it was on the job that the brothers changed the basic approach to shearing and became its fastest and most proficient exponents in the world. Traditional techniques had been adapted from blade-shearing days. The Bowens eliminated short blows and used long blows with the full width of the comb. Godfrey broke the world record in 1953 by shearing 456 full-woolled sheep in a traditional shearer's nine-hour day. The brothers each went on to break a string of records and were first and second in the inaugural national Golden Shears contest. They were hired by the New Zealand Wool Board to demonstrate and promote the new technique, with the emphasis on both speed and quality. Wool was glamorised as a by-product of their performances.

A consummate showman, Godfrey toured the world, became an international star and attracted enormous attention with his performances at the World Expo in Osaka, Japan, in 1970. A year later, he and a partner launched the Agrodome at Ngongataha, near Rotorua, a 160-hectare working farm, featuring farm tasks, including 1950s-style shearing, and performing farm animals, a tourist attraction that long survived him. Godfrey died in 1994 aged seventy-two.

Many skilled shearers had gone before the Bowens since cutting the fleece from sheep rapidly and efficiently had first become of economic importance to New Zealand in the 1840s. The first big-name champion

Above: The most successful competitive shearer New Zealand has ever produced is David Fagan (shown above, in a statue in his honour) who won his last Golden Shears title at the age of forty-four in 2006, his sixteenth New Zealand championship in twenty years.

Left: A competitor in a shearing event at the 2006 Royal Easter Show at Auckland.

was Charles Smith from southern Hawke's Bay who completed 200 sheep in a day with blade shears. Shearing machines came into general use in the early years of the twentieth century. Two Maori shearers, Johnny Hape and G. L. Stuart, both also from Hawke's Bay, finessed what had become known as a 'downward belly' technique in the 1920s. A number of other shearers improved their methods as the job became ever more competitive and a Maori shearer from the East Coast took the nine-hour-day record to 433 in 1934 – a record that stood for another two decades until the Bowens arrived on the scene.

The improvement of shearing speed and quality was of special importance to the industry because the problems besetting it were mainly marketing ones. The huge carpet-manufacturing business in the United States was baulking at the use of wool because of its unevenness of quality, with defects from a number of causes, including second cuts by shearers. Even more disconcerting were price fluctuations caused by the auction system which made forward planning difficult for carpet-makers and inclined them towards the synthetic fibres for which forward orders at fixed prices could be made for delivery on arranged dates. Also, New Zealand was floundering for wool-marketing skills it had never seriously needed before. It led to the great wool acquisition debate of the early seventies, with the grass-roots farmer, so to speak, resisting the attempt of the Featherston Street (political) farmers to buy in the national wool clip and control distribution and marketing in a way that had proved so successful in the dairy industry.

Top left: So much genetic material is now available internationally that sheep breeds new to this country are being taken up by enthusiasts. Damara sheep, available from East Asia, Egypt and South Africa, were imported in the new century. Damara is bred for its meat, does not need shearing, and, says the company, the meat has less than half the fat found in traditional New Zealand breeds. These sheep are on Tumeke Farms, Albany.

Top right: New Zealand agricultural tradition provided meat and milk from cows and wool and meat from sheep but over recent decades sheep and goats have contributed milk, mainly for cheeses. East Friesians, being milked here by Miles King on his Masterton farm, are big fertile ewes who can nurture up to five large lambs, and they provide milk for feta and other cheeses.

Opposite: Mustering sheep in the South Island high country on a beautiful clear day.

It was a revolution. Lambing percentages had for years been locked in around a hundred per cent or a shade lower. The Texel and Finn were followed by East Friesians (also a northern Europe breed) in 1996 and Dorper, a breed of South African origin, in 2000. The East Friesian was not only fecund but a prolific milker, kept primarily as a milch animal in some countries. Other late introductions included the Gotland pelt and white-headed Marsh. The Oxford Down was reintroduced in 1990 after disappearing in this country following original importation in 1906.

The first attempts to bring in the highly fertile breeds twenty years earlier had ended in disaster when scrapie disease broke out in a quarantined flock and all the sheep were slaughtered. A fatal disease that affects the central nervous system of sheep and goats, scrapie was feared because, once in the national flock, it would be almost impossible to eradicate – certainly if it became endemic to feral goats, for example. Scrapie is carried by an agent smaller than the smallest known virus, and is even more feared nowadays than previously because it is a transmissible spongiform encephalopathy (TSE) disease, similar to bovine spongiform encephalopathy (BSE), popularly known as mad cow disease, which has a link with other species, including humans. No known transfer has occurred between scrapie-infected sheep and humans but all TSEs are now considered threatening. However, even before mad cow disease had seriously damaged the rural economy of Britain, scrapie was considered a huge threat to the New Zealand economy. It is entrenched in the United States, where effects on the export and domestic sheep industry have been so serious the administration there decided to embark on an eradication policy – one that will take years.

However, by the 1980s, technology had triumphed, with the desired breeds successfully imported into New Zealand either as genetic material or as quarantined animals, or both. The new infusions of genes resulted in a better than twenty-five per cent improved lambing percentage as well as heavier lambs.

The average lamb weight in the new century was climbing from about 12.5 kilograms to sixteen or even eighteen. Some farmers pushed for twenty-five but most accepted that the industry had developed a market for a distinct type of New Zealand lamb, ideal for cuts. Other countries, including Australia, and particularly the United States, produced larger animals. The common joke is that American sheep are bred large so they can fight off the coyotes. So some New Zealand meat companies began discounting carcasses where farmers had bred big lambs and held them just that little bit longer on grass to get added weight.

An influence on these decisions were improvements in chilled meat technology and in access to Europe for chilled as opposed to frozen meat. The voluntary restraint agreement with Europe (VRA) originally provided for more frozen than chilled meat, but that

The muster on a sheep station is always a big occasion and its extent is governed by the size and topography of the farm. Here is a 5000-sheep muster on Lagoon Hills Station, Martinborough.

changed after New Zealand applied pressure to get access to the European common border. After that, individual countries that wanted to keep the lamb out had more difficulty arguing against it.

Demand for the chilled cuts grew and that changed the profile of the consumer – from the poorer end which bought cheap frozen lamb to the more affluent which preferred the chilled cuts.

The American market was similar, with no restraint on volumes since the aberrant bid during the Clinton presidency to impose quotas – a move blocked by the World Trade Organization. The Americans, too, had begun by buying New Zealand frozen lamb, and cooks were disappointed when they put it into the oven and it didn't emerge the same as fresh meat. The New Zealand industry researched ageing and conditioning processes, but chilled meat gradually became the prime export. The American market had always needed vigilance. For example, Americans enjoy lamb hot but seldom eat it cold, which gives an advantage to cuts rather than roast joints. It was also price sensitive. Once the price exceeded a certain limit, restaurant meals involving racks fell away, although some restaurateurs moved to lamb shanks and found them popular.

By 2005 sheep and beef cattle farms numbered around 13,000, most of them operated by families on basic grazing systems which depended in particular on the climate and the nature of the land on which they were situated. Most were on hill country which ranged from moderate to steep. The grass was supplemented by hay, silage, and in some cases fodder crops. Labour was provided by the farmers themselves, some casual labour and, increasingly, contractors. A crosscut profile of the average sheep/beef farmer in the early years of the twenty-first century was a man fifty-plus years old with 1500 ewes and 150 beef breeding cows. Mobile phones kept everyone in touch. The farmer had a computer but he seldom used it, whereas his female partner spent a lot of time online on long-distance purchases, price-checking and searching for good deals. The big need was broadband to provide the business with fast, up-to-date information. The need was urgent because these farms contributed more than twenty per cent of the value of New Zealand exports.

CHAPTER SIX

THE VAGARIES OF BEEF

THE PREVAILING ANGST among New Zealand farmers in the 1960s and 1970s was that foot-and-mouth disease would arrive and destroy an economy that remained so dependent on the export of pastoral products. But, at the same time, they wanted to find a risk-free way to bring in other cattle to spice up beef and dairy breeds.

Remoteness and long, slow journeys to get here had been historical allies against many exotic diseases. But direct contacts with the outside world had grown in volume and speed. The first passenger jets brought in bigger numbers of people and freight more swiftly and regularly than ever before from the early 1960s. The tempo increased further a decade later with the daily arrival of jumbo jets carrying up to 400 people and large quantities of cargo.

Farmers and the Department of Agriculture – although wary of a number of diseases historically unknown here – most feared foot-and-mouth, a highly contagious viral infection with low mortality in adult animals, although it may kill the young. Foot-and-mouth can spread through bovine species, sheep, goats, pigs and deer at a frightening pace. The incubation period is up to a fortnight, and if infection were to spread into feral animals, eradication from New Zealand would be a logistical nightmare.

The slaughter of infected stock and all stock that had come into contact with victims of the disease would follow an outbreak. All stock movements would cease, and no other country would take pastoral exports until New Zealand was declared free of the disease. Because of the possibility of reinfection, all slaughtered animals and any associated animal products would be burnt, along with any farm material or equipment that could not be scrubbed and disinfected. Foot-and-mouth remains endemic in many Asian, African, Middle Eastern and South American countries, and outbreaks are a constant threat to Britain and Europe, mainly through the smuggling of illegal meat. New Zealand government departments trained many people – from veterinarians and farm advisory officers to police and customs officers – during the three decades after World War Two, first, to keep out foot and mouth and, second, to move swiftly to eradicate it before it could spread should it arrive.

But to overcome this dread of disease arriving with imported animals, techniques were developing

Contagious diseases such as foot-and-mouth and mad cow disease have had serious economic consequences for British farmers. Not only infected animals but animals that have had any likely contact with the infection are slaughtered and burnt, like this victim of mad cow disease.

Among the many exotic cattle breeds imported in the 1970s and 1980s were the big-framed Swiss Simmentals (above), which quickly gained supporters, and Highland cattle (left), which were as much a curiosity as an economic contributor to the beef industry. The Highland breed is able to survive in tough conditions.

as early as the 1960s to enable beef breeds to be brought in, even though the difficulty of detecting scrapie stalled bids to bring in new sheep strains. The Dairy Board, through Livestock Improvement Association, was enthusiastically involved in devising ways of safely importing new cattle breeds.

The first beef animal in New Zealand was the Durham, an early type of shorthorn, which arrived from Australia with missionaries in the second decade of the nineteenth century. Later-type shorthorns were imported in 1849 and became popular because of their all-purpose role as a draught, dairy and beef animal, which made them especially useful in remote areas where settlers were farming not far above subsistence level. The shorthorn was superseded as a dairy cow by the single-purpose Ayrshire and Jersey but was present among beef herds for many decades. During the nineteenth century, the Hereford and the Aberdeen-Angus (as it was called then) gradually took over the beef production role, being ideally suited to both New Zealand's fertile lowlands and steep hill country. Of the two, the Angus gained ascendancy, perhaps because it was better on the hills, able to survive comfortably and produce calves that put on weight while still on the slopes. But both breeds are hardy and able to recover quickly from periods when grass is scarce. Herds in the modern beef-fattening industry are mostly made up of cross-breds from Angus and Hereford. It gets harder as the years go by to find herds of pure-bred Angus and Hereford breeding cows.

For all their anxiety about exotic diseases, farmers believed that importing new exotic cattle breeds was important for the future development of the pastoral industry in this country. New genes might expand the context of farming both by producing bigger and more fertile sheep and meatier prime beef animals. One of the big moves in the 1960s was into dairy beef, using beef bulls as terminal sires over dairy cows which were surplus to requirements for dairy herd replacement. Large-framed European breeds might prove ideal for that job, especially with the growing spread through dairy breeds of the Friesian, a large-framed dairy animal good enough on its own for beef and ideal for crossing with beef sires.

Of the several exotic breeds brought in since the 1960s, most have been used, and very effectively, as terminal sires over dairy cows. The upgrading of dairy beef quality with the use of beef sires became a major campaign in the late 1960s and the use of

A champion shorthorn bull, 'Glenburn Keepsake', paraded for the camera, probably by his owner, Mr D. J. Holden, in 1947.

new genes was part of it. British cattle had been developed for prime, fat carcasses, whereas European animals had been bred for pulling wagons or ploughs, and their large, heavily muscled frames provided lean meat.

The first of these Continental cattle to arrive in New Zealand – and it did so with a fanfare – was the Charolais, the big-rumped white animal that originated in France but came here first from Britain in the form of semen in 1965. Sixty-one bulls followed between 1969 and 1981. Calving problems emerged early on but these were gradually ameliorated. Then the gene pool became flooded with a tsunami of breeds from many parts of the world.

After the Charolais came the Maine-Anjou, Limousin, Saler and Blonde d'Aquitaine, all also from France; Simmental, a big-framed Swiss breed; Chianina, the tallest cattle in the world, Romagnola, Piedmontese and Marchigiana from Italy; Santa Gertrudis, a breed developed in Texas from shorthorns and Brahmans; and Murray Grey, a breed fixed from shorthorn and Angus in Australia. In 1990 Shaver cattle were introduced from Canada, where the composite animal was developed from nine breeds. Surprisingly, none of these animals butted out the traditional Angus and Hereford, and some remain nothing much more than curiosities. Yet Charolais and Limousin are present in numbers and some of the others have their fans. The role of these exotics became mainly that of terminal sires for dairy cows, and no doubts lingered at the beginning of the twenty-first century over their contribution to dairy beef production. By then, dairy beef made up well over half of beef exports, mostly to the United States for manufacturing into hamburgers.

But none of the introduced breeds challenged the ubiquity on the hills and the fattening paddocks of the Angus and Hereford. These cattle had adapted well to the New Zealand environment and the farming practice of outside grazing that had prevailed over the years. Just over twenty per cent of national beef production was from the Angus, and about thirty per cent either Hereford or Angus-Hereford cross. The rest was dairy beef from Friesians or from crosses usually involving Friesians.

The stocky Angus arrived here first in 1863, six years before the first Hereford, and within a few years farmers were exporting the cattle to Australia. Although initially the Hereford was more favoured, the Angus caught up in the late 1920s and has since surpassed all other beef breeds. In the past few decades, Angus cattle and semen have been brought in from Britain and North America to refresh the locally adapted animal, which has itself been exported to countries in North America, Asia and Europe, and even back to Scotland where the breed originated.

Beef cattle were historically regarded as ancillary to sheep farming. They were used for pasture control on hills, trampling and eating rank grasses and shoots of fern, manuka and other woody second growth, especially on land that was not carrying enough sheep to economically warrant the regular application of fertiliser. As one scientist wrote in the mid-1930s: 'A long and arduous experience in hill country management has . . . resulted in the utilisation of cattle as the best machine for pasture and second-growth control.' Some farmers believed sheep prospered better sharing grazing with cattle not only because they ate the coarse growth but because they also ingested the larvae of internal parasites which infected sheep but failed to gain accommodation in cattle. Sheep farming and dairying were solidly profitable in the 1960s, but

An Angus cow and calf in Tuki Tuki, Hawke's Bay. The Angus, originally the Aberdeen-Angus, has been traditionally the most popular beef animal in New Zealand, just ahead of the Hereford. It adapted superbly to farming systems here.

fattening specialised breeds for beef was not an enterprise that could exist economically on its own; so beef was essentially a by-product of sheep farming and dairying. The specialised beef breeds were most densely farmed on high country sheep properties in Hawke's Bay, Poverty Bay and the Wellington region.

New Zealand has never seriously contested the world market for prime beef which comes from cattle finished with grain on feeding lots thus developing the 'marbling' of intramuscular fat that enriches flavour and has always commanded a premium in America, other traditional Western markets and in Japan. Australia, with cheap grain available, grasped some of this prime beef trade.

The top New Zealand beef, free-ranging and grass-fed, comes in below that and will continue to do so in the foreseeable future.

However, the farming of cattle for beef altered over the last decade of the twentieth century and became economically more important as markets opened wider in the inaugural years of the twenty-first. Breeding cows, about 1.5 million of them, still mainly grazed hill country and remained effective at improving pasture quality. Mostly delivering a ninety-five per cent calving rate, they weaned calves at 230 kilograms' live weight or better. In a growing number of cases, the young cattle were finished not on the hill country farms but on flat and fertile properties whose owners exploited the difference

Raines shine

Back in 1844 John Waring Saxton bought the unwanted, prefabricated New Zealand Company's barracks at Port Nelson and moved it to nearby Stoke, where it became 'Oaklands', the homestead on his 2000 acres of sheep and cattle country. Eight generations on, Saxton's descendants still live in the house and still farm the property.

John Saxton landed in Nelson in May 1842 with his wife Priscilla, their five young children, his mother-in-law Mrs Crumpton, and his brother and sister-in-law. The property at Stoke was difficult to farm, comprising mostly hills with swampy flats that ran down to the coast. Saxton had no farming background but he got by. A decade later he became the first lessee of Tarndale Station, hard country between the head of the Clarence and Wairau valleys, and stocked it with cattle.

John Saxton became involved in Nelson's civic affairs and was noted as a gifted watercolourist and a musician, but as the generations went by the farm fell away economically. In the 1920s Englishman Richard Raine arrived. He had been to agricultural college and worked on an uncle's farm before moving to Western Australia. After two years there, he moved to Nelson for a look at New Zealand. He met and courted Gwen Saxton, and they were married in 1927.

The farm was carrying sheep and a few beef cattle, but it was heavily in debt. Much of the land still needed to be cleared. Some was still in scrub, and gorse had spread over the hills after it had been planted as a hedge to remind the early members of the Saxton family of England – the same impulse that led to many thousands of gorse hedges in New Zealand that subsequently broke out into the paddocks they were supposed to define. Moreover, the debt was compounding at eight per cent and the mortgagors were threatening to foreclose. When Richard and Gwen were married, the family asked him to take over the farm. He talked up a deal with the mortgagors and the family sold off much of the best flat land to protect the rest.

The family survived troublesome times. Eighteen months before

Since 1844 the family has lived in the same house and farmed the same land.

John Saxton moved his family to Oaklands in November 1844, Arthur Wakefield and forty-nine settlers had been killed while rashly trying to arrest Te Rauparaha and Rangihaeata in the nearby Wairau Valley. The house still has a small attic called the 'Maori room' in which the children were to hide in the event of an attack. You can still see the windows, painted out.

Richard moved into dairying and became one of the people involved in setting up the town milk industry on a national basis in 1944. Previously, throughout the country, dairy farmers had distributed milk locally, ladling it from can to billy. Worries had been spreading about the unsupervised quality of the milk and its potential to spread disease, and also about availability and price. As the result of a report from a between buying price off the hills and the later selling price. Most animals were taken through for slaughter at twenty months and around a 300-kilogram carcass. How much money the finishing farmer made depended on the cost of feed and the prices on the North American market.

The totality of beef cattle, including surplus from the dairy herd, numbered more than 4.5 million. The biggest concentrations were in the Waikato, Manawatu-Wanganui, Hawke's Bay, Canterbury and Northland. Although New Zealanders ate about twenty per cent of the beef produced (as always, more per capita than lamb), export prices remained heavily dependent on the American market, which in turn was affected annually by the level of domestic beef production and by the availability and price of pork and poultry. However, producing beef on its own had become a viable choice, given the expertise and the right sort of land.

The industry was helped, as lamb exports had been, by the improvement in chilling techniques, providing a much more attractive product than freezing. Chilled beef was first exported to Britain in the 1930s but the distance was great and handling distribution, obviously more urgent than for frozen meat, was a constant problem. But, with some prescience, an agricultural economics lecturer at Canterbury Agricultural College, Lincoln, I. W. Weston, wrote in 1935, 'Any major expansion of

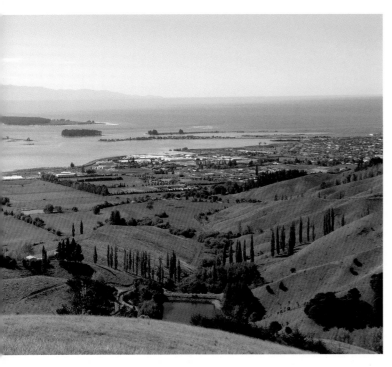

government-appointed Milk Commission, forty-four authorities were set up around the country to mollify consumer anxieties. So from 1944 onwards milk was pasteurised and bottled and farmers were contracted to supply quotas. The Raines began under the new regime with fifty cows – a big herd then – farmed on the foothills of their property. By the end of the century they had a herd of between 250 and 300 Friesians, of which part calved in autumn and the rest in spring to supply a milk quota year-round. They also bred and fattened their own dairy beef and grew maize for supplementary feed.

In the 1960s and 1970s Richard Raine's son, also Richard, bought back three of the blocks sold off when the farm had been under economic duress thirty years previously, but what had become Saxton Fields, a complex of sports grounds, and an industrial area on the coast, was unobtainable. The cows remained the farm's fundamental activity but the family enterprise responded brilliantly to the need for change in the late 1980s. On the main property, and on land owned or leased nearby, the Raines branched into forestry and horticulture: apples, kiwifruit, boysenberries, blackcurrants and hops.

The family's tie to the land has deepened over the years. Richard Raine has lived in the house since he was born in 1936, and his emotional attachment to both the house and the farm has grown with the years. If it were possible – and it really was not – he would have liked to restore some of the flats into wetlands. The house, registered with the Historic Places Trust, was kept in first-rate condition with the original Scottish pine weatherboards held on with hand-cut nails. A room was added, designed to be in keeping with the original style. The garden remains cared for with devotion. As urban Nelson crept closer and the rural differential was abandoned, rates rose and it would have been easy for the family to make a fortune by subdividing. 'Yes, but I could only do it once, and this is our heritage,' said Richard Raine. 'It would be like selling the family silver.'

His son, Julian, said that many people have their treasures, their heirlooms. Their own family heirloom after more than sixteen decades was, he insisted, Oaklands – the house and the land. His was the seventh generation to be born in the house and he later lived in a home of his own less than a hundred metres away. A former Nuffield Scholar, and horticulturist and businessman, Julian drove the diversification of the family enterprise. He felt, he said, that the family had been there so long they were tangata whenua, and Ngai Tahu seemed to agree. They elected him chairman of Ngai Tahu Horticulture, which administered the collected Maori horticultural land in the region. As well as that he was on other boards, including chairman of Berryfruit Exports New Zealand Ltd which exported about ninety-seven per cent of boysenberries from New Zealand.

Since Richard and Gwen Raine took over in the 1920s the farm has thrived, emerging into the twenty-first century as a modern, entrepre-

Above: When carcasses were manhandled in and out of the freezer at the Gear Meat Preserving and Freezing Company in Petone, staff were padded up against the severe cold. Nowadays, freezing and chilling facilities are automated.

Right: Two Gear labels from the 1880s. In those early days of the export industry, consumers regularly enjoyed eating many parts of the animals, including, as advertised here, ox-cheek, stewed kidneys, potted head, tripe, ox and sheeps' tongue and pigs' feet.

the beef industry in this country is likely to depend on the success with which chilled beef can be marketed overseas. At present the marketing of chilled New Zealand beef is in the early experimental phase and its ultimate outcome cannot be foreseen.'

Too few beef cattle complementing sheep was then seen as a problem, and Weston continued that more attractive returns from beef would result in 'An improvement in sheep carrying qualities of the pastures . . . when the roughage was properly eaten out. In such circumstances beef cattle would rank not only as mowing-machines but as direct income-earners as well and the farmer would, consequently, have a wider field over which to spread his risks.'

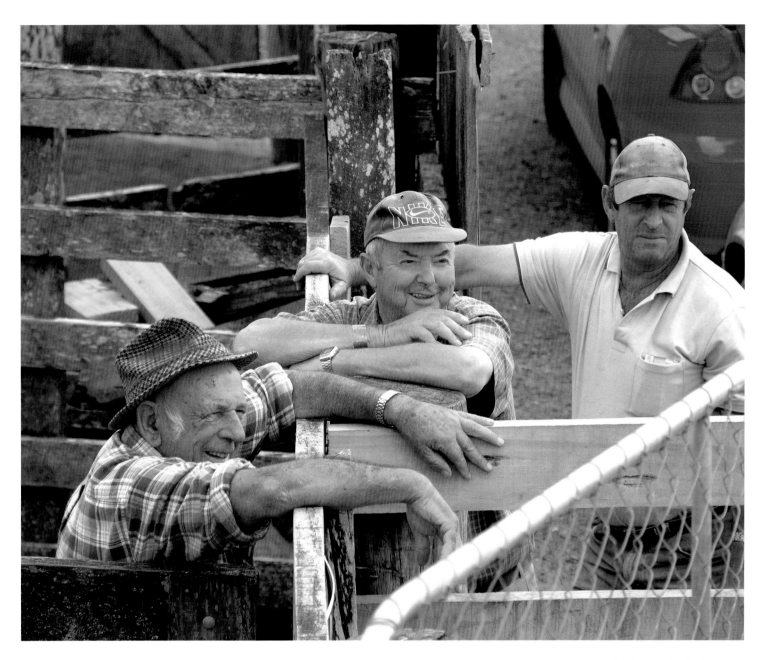

In the opening years of the twenty-first century, negotiators gained two advantages as developed countries held out against free entry. Quotas of beef and veal into the United States were increased but still carried a tariff. Beef export quotas to the European Union were increased threefold, although only to 1300 tonnes and at a twenty per cent tariff. In the meantime, New Zealand beef farmers were exploiting every other opportunity. As dairy herds built up spectacularly after the late 1980s, some farmers began holding on to all their dairy calves and running them on leased hill country for sale as milking cows, and some beef farmers bought heifers in to run on their hills for resale later to farmers building up their dairy herds. Friesian bulls from dairy herds, fast-growing and with lean meat, also became available in huge numbers, and in some parts of the country replaced Angus-Hereford cattle on the rolling hill country and even on steep hills that the beef breeds had called their own for so long.

Beef, like most New Zealand pastoral products, faced the long-term difficulty of providing a product for developed countries with Western tastes – the very markets which financially supported pastoral industries of their own and erected barriers to keep cheaper suppliers out. New Zealand's share of world beef trade was not as great as with lamb and dairy products. Australia and a number of South American countries also produced large quantities, crowding the available international markets.

Saleday, the day the farmers came to town, was the social event of the week in hundreds of New Zealand towns for more than a century. Animals were driven along country roads or later trucked to the saleyards where the business of buying and selling stock took place. The stand-on-the-rail auction has less influence on rural life and business now but the culture continues.

CHAPTER SEVEN

DEER AND GOATS

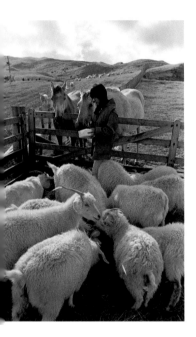

Angora goats, from which the coveted angora fibre is shorn twice a year, are mostly farmed in the drier regions of the country.

In the space of twenty years from the 1960s, deer were transformed from feral pests to domesticated farm animals in an extraordinary demonstration of the economic adaptability of the rural New Zealander. In many countries deer had been farmed, in the loosest sense of the word, for centuries – by nomads who followed migrating herds and lived off them, by Asians who kept them in enclosures, and by Europeans who confined them in forest parks for the aristocracy to 'hunt'. But in this country they were incorporated within a modern, intensive, grasslands farming system for the first time anywhere in the world.

The speed with which deer were plucked from the bush, placed in paddocks behind high fences and domesticated to the degree that they could be controlled by farmers astonished many. And yet the impetus for the initiative was simple enough. An infrastructure of packing houses had been set up to service already developed export markets for venison created by hunters increasingly using helicopters to shoot deer in the bush and bring out the carcasses for processing. So great was their success that feral deer numbers slumped and the surviving animals took to more and more remote country. With supply drying up, the simple solution was to capture deer, breed them on farms and provide a base for the industry.

The pioneers of the deer-farming industry, led by Sir Peter Elworthy, were held up to ridicule – deer were wild animals and would never be domesticated. But even the farmers themselves were astonished at how, within just one generation, deer took to farm conditions. Two things were important to the establishment of a secure industry. The first was that two markets already existed: one in central Europe for venison; and the other in Asian countries, most particularly Korea, for deer antlers, which fetched high prices as an important ingredient of Asian medicines. Both these markets welcomed the prospect of increased supplies of product. The other factor was the determination of a few farmers to ignore the red tape that hampered the other primary industries and to find out for themselves what could and could not be done.

Red deer were liberated in New Zealand by British settlers from the 1850s. The idea was to provide game for hunters, and the big 'heads' that soon became common in this country lured wealthy British aristocrats to Otago and the Wairarapa, providing employment for several former Scottish gamekeepers. Deer increased in such numbers, however, that the system of licences being issued by then ubiquitous 'acclimatisation societies' was dropped and deer shooting became an egalitarian activity in New Zealand. New breeds followed red deer (*Cervus elaphus*): fallow, sika, sambar, white-tail, rusa and wapiti, as well as moose and gnus (African antelopes), neither of which survived. Those that did survive thrived in the congenial climate. Feed was abundant and the only predators were human beings with rifles, whose preference for stags with 'big heads' did little to lower deer numbers. And hunters were relatively few in a largely empty country anyway. However, by the first decade of the twentieth century, deer

Farmers were astonished how short a time it took for deer to become domesticated farm animals, grazing behind the high fences their high-jumping skills made essential. These deer are on a farm near Rotorua.

had become an ecological menace, causing large-scale erosion by stripping native grasses and shrubs from the flats and slopes of watershed valleys, and by eating leaves and shrubs, causing bush to degenerate. Numbers had increased to the point that they seriously lowered sheep stocking rates on the high country and led to farmers calling for government action. Their complaints carried weight because wool was still important to the New Zealand economy. The situation was regarded as so serious that a 'Deer Menace' conference was held in Christchurch in 1930 and for the first time all protection was removed from the animal. The Internal Affairs Department mounted a campaign against what were declared 'noxious animals' and the task of dealing with them was given to G. F. Yerex, who became known for the rest of his life to all professional deer shooters as 'The Skipper'. Yerex recruited men used to the backblocks, especially high-country shepherds and hunting guides.

These hardy men became known in later years as 'cullers' – a misnomer given that their task was eradication not culling. They shot hundreds of thousands of deer, and left their carcasses to rot in the bush but were encouraged to take tails, pizzles and sinews to supply the Asian market. The department's primary aim was to kill as many deer as possible, but the hunters were well aware of the value of the skins, antlers' various by-products, and even the venison if it could be got to the chiller in time. Such facilities were springing up in very remote areas and many government hunters took up deerstalking as a profitable enterprise. In fact, the profits were so good that jet boats and light aircraft become involved. The post-war deer culler was romanticised by author Barry Crump with his bestselling *A Good Keen Man*, published in 1960, and later novels.

The export market in Germany expanded steadily and good prices led to an increase in the infrastruc-

Above left: The young Barry Crump with 'Flynn' (front), the dog that figured in his bestselling book, *A Good Keen Man*, which romanticised the life of the post-war deer culler.

Above right: The life of helicopter pilots and shooters was hazardous and this became reflected in insurance premiums.

ture. The Vietnam War with its gunships inspired some entrepreneurs to use helicopters and hunters took to the air, shooting deer with automatic rifles, retrieving the carcasses, delivering them to the nearest roadside point and then returning to the slaughter. The government was at last seeing deer numbers plummet at no cost to the taxpayer, and an export business was flourishing into the bargain. In their own way, the participants were freebooters, ignoring rules and regulations, and several helicopters paid the price – especially as competition among them heightened. With deer numbers dropping so fast it became hard enough to cover the cost of operating a helicopter let alone climbing insurance premiums. But when individuals realised the future lay in farming deer to improve availability of the produce at the right time for the market,

the helicopters came into their own again, this time to bring back live animals in nets.

Businessmen who had financed the take-off of the export industry, together with some enterprising shooters and farmers, then looked at the prospect of creating an orderly supply, putting captured deer behind the high-cost, two-metre-high, wire-netting fences required and breeding them. In the 1960s, when some deer were already behind fences, pressure went on the government to formally authorise the farming. Acting on the advice of a select committee of Parliament, the government came out in favour, despite opposition from some farming and environmental groups concerned about the possibility that animals escaping from the farms would restock the feral population and again threaten the native forest. The government's decision to approve the new industry was driven by an accepted need at that time for diversification of products and markets. The decision proved the right one, with deer farming swiftly becoming the largest new pastoral enterprise of the century.

Controlling regulations were constrictive at first but loosened as farming proved successful and associated problems minimal. Initially the deer were run on poor tussockland because that had seemed their habitat of choice in the wild. It didn't take farmers long, though, to discover that the deer had lived in rough country because they were safest there from hunters – that, as with sheep and dairy cattle, they thrived best on good, open pastureland with, of course, silage, hay or grain supplements when required in winter. No one anywhere had devised systems for running deer on grass in modern pastoral farming conditions so many farmers treated the farm as an extension of the bush. They killed their stags by shooting at between fifteen and eighteen months but the New Zealand reds had not reached target market weights by that time.

Experience and common sense triumphed as practical management systems were devised, aided by research from the scientists at Invermay research station in Otago. Fence and yard designs, feed content and patterns, mating management, genetic selection recording, optimum economic weights for slaughter and other issues deeply understood by sheep and cattle farmers had to be worked out anew for deer. Most farmers stocked with red deer but, to achieve market weights in time for the European Christmas trade, elk (wapiti) were imported from Europe to produce yearling carcass and antler weights. In the late 1990s stags with massive antlers were still being imported from Europe and Russia. The locally captured feral deer were gradually replaced by farm-bred stock which proved more tractable, although stags remained temperamental during the roar.

The government encouraged investment in the new enterprise as it was being established, which resulted in millions of dollars from city investors pouring in. Farmers throughout the country were cautious at first, but once the returns from deer were compared with those from sheep and cattle there was a rush of interest. Other countries also displayed interest in what was happening and adopted New Zealand systems, but none had the high numbers of feral animals with which to build herds rapidly.

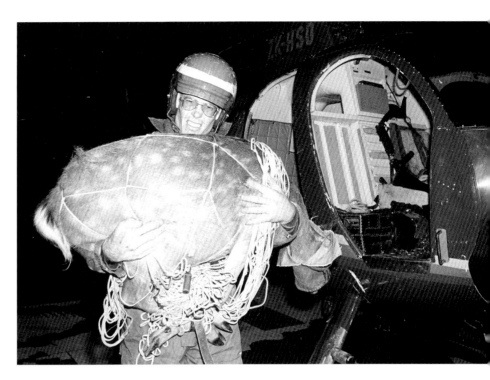

A live hind captured in a net and strung up for delivery to a deer farm was worth as much as $2000 to the captors.

Several firms were exporting venison but were persuaded to co-operate in marketing. This led to the formation in the mid-1980s of the Game Industry Board (succeeded by Deer Industry New Zealand). In 1992 venison was rebranded as Cervena in a bid to clearly identify the New Zealand product, and promotion in the high-value European market was stepped up, especially in Germany, with its traditional affection for the meat, and Scandinavia. The much-vaunted analogy was that the especially tender Cervena from open New Zealand pastures was to venison what Champagne was to sparkling wine. The name, the promotional literature declared, came from a combination of cervidae ('deer') and venison (which originally meant 'hunting'). For a while, the new name was heavily promoted within New Zealand in the hope of widening venison's acceptance domestically, especially as a healthy lean red meat.

While Germany, and a number of other countries, had followed the New Zealand example and begun farming deer, its total of animals as the old century turned into the new was about one-twentieth of the national New Zealand herd, which numbered around two million – about the same as the combined total of all the other fourteen countries

The New Zealand deer industry branded its venison as Cervena to get a distinctive place in the market. David Kratz was involved as a public relations consultant in the launch of Cervena in the United States in 1993 and is pictured here (left) with Minister of Agriculture John Falloon when he was keynote speaker at the 1996 Deer Industry Conference.

farming deer. By the time the industry had settled down, herds numbered more than 4000, many run as a secondary enterprise to sheep or cattle. Some specialised in venison while other farmers raised stags with antler size and formation that brought high prices. The farms were spread around the country but were most common in Southland, Canterbury and the Bay of Plenty. The main supplier of velvet internationally was Russia but New Zealand became regarded as the second best source.

However, the industry in New Zealand gained too much momentum in times of strong earnings, with hectares carrying deer jumping more than sixty per cent in the decade from 1994. The growth of production that went with that expansion had a major impact on world prices for venison, with exports jumping from 22,700 tonnes in 2004 to 27,431 tonnes in a single year. Most of this increase went to Germany but slightly increasing volumes went to France and Sweden, and lower-value manufacturing meat went to Australia and Taiwan. Marketers were aiming at the United States as prices fell amidst the growing volume of venison on sale in Europe. Deer farming profitability in New Zealand was in decline because of overproduction, and there were growing signs that some farmers were reducing their involvement.

The other product, deer velvet, was also coming under price pressure in its main markets in Asia, particularly Korea. Velvet is the antler tissue on male deer, harvested before it hardens and calcifies.

Depriving stags of their fighting antlers also helped manage them on the farm. The velvet renews itself every year and has a traditional market in Asia as an aphrodisiac and as an ingredient in a range of health products. The volume in world trade was affected by the fact that velvet harvesting was illegal in some countries, mostly in Europe, for animal welfare reasons. In New Zealand the tissue was taken using anaesthetics under veterinarian supervision.

By far the biggest buyer of New Zealand velvet was Korea, with varying levels of demand from China, Taiwan and Hong Kong. The market proved erratic, depending on fluctuating volumes of local supply, on the effect of Asian economic strength on discretionary spending, and on trade protection. In a bid to increase the demand for velvet, New Zealand producers, in association with AgResearch at Invermay, supported investigations into velvet as a dietary supplement promoting good health, enhanced athletic performance, sexual energy and acting as an agent for anti-ageing and immune system improvement. Claims, some extravagant, were being made before research had advanced very far at all.

Farming another introduced pest, goats, never reached the same level of economic importance as deer because no market developed that approached the value of venison demand in Europe. Although more goat meat was consumed in the world than sheep meat, almost all of it sold in Third World markets where returns have always been insufficient to attract New Zealand producers. Thus, goat meat and dairy production remained a small segment of New Zealand's primary industry and was never able to shift a significant number of domestic consumers from strongly traditional wool, lamb, beef and bovine dairy foods and, therefore, was unable to develop an industry large enough to gain economic benefits of scale. Goats enjoyed something of a boom during the sometimes frantic moves towards diversification in the 1970s and 1980s, but the industry was disorganised and fragmented, and only in the 1990s was it gaining a small but more rational place among the primary industries.

Goats were introduced to New Zealand by Captain Cook and were later brought in in large numbers by whalers, sealers and the earliest settlers

A deer with the velvet, or antler tissue, which renews itself every year and is a valuable export to Asian countries, notably Korea.

during the closing years of the eighteenth century and the beginning of the nineteenth. The intention was to domesticate some for their meat and to release others for hunting, not only on the mainland but as emergency food on many of the islands visited by whalers and sealers. However, wild goats in their hundreds of thousands proved especially harmful to native bush, and Department of Conservation hunters have shot vast numbers in a bid to keep populations down. With their hard mouths and voracity, goats proved useful also for keeping noxious weeds, especially blackberry, at bay on marginal land, and are still used for weed control.

Goats have been domesticated farm animals for thousands of years and remain predominant producers of fibre, as well as of meat and milk in many parts of Africa and Asia. Both the meat and milk products are of proven nutritional value. Goat cheeses have substantial markets in Europe, especially in France and the Netherlands. But the advantage dairy cows and beef cattle enjoy over goats is their greater productivity.

Commercially, the most successful goat farming has been fibre: mohair, the fleece from Angoras; cashmere, the fine underdown grown by all goats (except Angoras); and cashgora, the fleece from Angora/feral crosses. Mohair is shorn twice a year from Angora herds mostly sited in the drier regions of the country. Most of the fleece is sold raw overseas where it is sorted and processed into products

Visitors with Saanen dairy goats at Mike and Ruth Redwood's farm near Geraldine. Goat's milk gained a wider acceptance in New Zealand during the last quarter of the twentieth century as the industry organised itself under the aegis of Dairy Goat Co-operative (NZ) Ltd. The farm supplies milk to Talbot Forest cheeses.

for the high end of the clothing market.

Angora fleece is similar in composition to wool but more elastic with greater lustre and takes dye well. Its end use is the fashion industry as well as luxury furniture fabrics. The industry in New Zealand has had its small booms but too many busts, mainly because it was based on inferior breeding stock. Most of the world's supply is grown in South Africa and the United States (particularly in Texas), and new genetic material brought in from these two sources in the 1990s promised a more solid future for the industry based on improved quality.

China is the world's leading producer of cashmere, a yarn that also has its place at the luxury end of the world fibre market for clothing. Cashgora is a New Zealand product, a coarser fibre than mohair, which has found its own place in the textile industry.

Milking goats have been farmed in New Zealand for many years and gained a wider acceptance during the last quarter of the twentieth century, although the industry survived several periods of overproduction that forced some producers out of business and financially damaged others. The farmer-owned Dairy Goat Co-operative (NZ) Ltd gradually became the dominant producer and processor of dairy products, and like the bovine dairy industry was centred on the Waikato. In the 1980s the co-operative developed what it claimed was the world's first infant formula based on goat's milk. Most of New Zealand's goat's milk is processed into powder and manufactured into consumer products, such as UHT milk, and exported, with heavy marketing emphasis on nutritional value. It is an industry that became well organised and developed a stable base for future growth.

Goat carcass exports number about 135,000 a year, more than eighty per cent of them from feral or semi-feral animals mustered twice a year for slaughter. This represents around eight per cent of the world trade. Few farmers have concentrated on breeding up for meat with such a small market in New Zealand, but the importation in the 1990s of the Boer breed from South Africa brought some enthusiasm to the goat meat business. The Boer is a high-fertility, early maturing, specialist meat breed, with faster growth rates than the crosses already in the country and produced good results from hybrid vigour when bred with local animals.

CHAPTER EIGHT

THE FALL AND RISE OF FORESTS

One of the major primary industries developed in New Zealand since World War Two processes timber from a forestry resource that began in reafforestation schemes half a century before.

In the earliest days of European settlement, the indigenous timber resource seemed limitless and was the raw material for one of the first major extractive industries – along with kauri gum, the resin sap in the ground under the great kauri stands. Kauri grew only north of a line from Port Waikato to Tauranga but the Northland forests covered the largest area. Until well into the twentieth century gangs of diggers operating in Northland – many of them Dalmatians (mostly then called 'Austrians') – worked over the land, probing and digging for the gum, which was valued for its use in making varnishes and polishes. Maori themselves had long used the gum as fuel and for torches, and Captain Cook found some on the beach at Mercury Bay but thought it came from mangrove trees. The first exports went to Sydney in 1815, and by the 1830s its value was established and Bay of Islands and Auckland merchants were purchasing it from Maori to send away. Farmers in Northland dug for the gum under the old kauri forests to supplement their income and the Dalmatians began arriving in 1885. At the turn of the twentieth century, gum exports were earning an average of half a million pounds a year – about the same as flax and a lot more than cheese.

With its tall, straight, branchless trunks rising into the sky, kauri provided timber that was also exported in quantity from the earliest days because of its value in shipbuilding. However, one of the most valuable of the native timbers was kahikatea, or white pine. Kahikatea has a straight grain and long clean lengths. Because it was odourless and therefore did not taint food products, it was widely used for butter boxes, cheese crates and other food packaging. Until World War Two it was by far the most commonly exported timber, with rimu, valued for furniture, second. By the end of the war kauri exports had stopped, and indigenous timbers ceased to be of export value.

Timber production from indigenous forests peaked in 1907 at 432 million superficial feet, spectacularly up from 261 million feet in 1900. Exports kept going up, though – until 1912, when ninety-five million feet were shipped out. Imports of timber – mainly Australian hardwoods for railway sleepers and telegraph poles – boomed during the first decade of the twentieth century to fifty million feet.

Estimates are that between eighty and ninety per cent of the country was covered by forest when Maori arrived and that that area had been halved by the time New Zealand became a colony in 1840. Maori fired large areas – often larger than intended, by accident – to establish gardens and to flush out the abundant flightless avian game. Although the forest provided some fruits and other food for Maori, it was far from the richest source. The deforestation dramatically reduced the number of indigenous animals and also whole species. When Europeans arrived, the tempo of burning for gardening by Maori increased in order to grow

The Fall and Rise of Forests 229

Gumdiggers in North Auckland in 1914 show off a week's work when kauri gum was still a major export earner.

Digging gum was not enough for these men, in 1910 near Kaitaia, who bled living trees for their precious sap.

potatoes and other crops to supply ships and settlers; and settlers then took over the burning in a much bigger way to establish pasture.

Alarm bells at the scale of deforestation were sounded from the 1870s. Authorities were becoming alarmed by the rate at which the forest cover was being slashed and burnt, especially in the North Island. In 1871 some provincial governments offered individual property owners four pounds for every acre planted with any kind of tree. In Canterbury and Otago strong north-west winds made large shelter belts attractive and many took advantage of the offer to plant radiata pine and macrocarpa plantations, as well as smaller areas in blue gum varieties. The Railways Department established small plantations of exotic trees during the 1880s.

The big moves began in 1896 when an afforestation branch of the Lands Department was formed and forest tree nurseries were set up at Tapanui and Eweburn in the South Island and Rotorua in the North. The favoured tree became the exotic from North America, radiata pine, because of its extraordinarily rapid growth in New Zealand conditions, maturing in not much more than twenty-five years – as fast as anywhere in the world. Macrocarpa remained a popular tree for single-line windbreaks.

The first major planting began in 1898, and ten years later plantings had reached a thousand acres a year. From the 1920s until 1936 came the planting surge (aided by relief labour from the unemployed during the Depression) that covered another 270,000 hectares – 150,000 by the state and 120,000 by private companies. In 1927 alone 4.8 million trees were distributed from State Forest Service nurseries, mostly to individual private planters. Both state and private activity was involved. The largest plantings were in the central North Island. By the 1960s claims were proudly made that Kaingaroa was the largest man-made forest in the world.

In 1961, twenty-five years after the end of the planting boom in 1936, state planting of exotics was increased again and the government dangled incentives in front of private companies to extend their forests as well. By the end of the seventies, exotic forests covered about 750,000 hectares, mostly in the centre of the North Island, the site of the highest concentration of planting in the 1923–36 period. Generally, the availability of undeveloped land dictated where the forests were located. The plateau was a wise site for the province because the land in the region had for many years been considered marginal for pastoral farming because of the mystery ailment called bush sickness that eventually led to serious unthriftiness in stock. The discovery by researchers that it was caused by mineral deficiencies in the soil, particularly cobalt, brought many thousands of hectares back into pasture, although the area remained host to the biggest forests and planting continued on marginal land.

After World War Two, New Zealand Forest Products Ltd (then the country's biggest company) and Tasman Pulp and Paper Co (formed in 1952) built an industrial superstructure over the burgeoning production of exotic timber from the maturing

By the 1930s, when cheap labour became available during the Depression, teams were employed to plant exotic pine trees across the central North Island plateau, eventually creating Kaingaroa Forest which, it was claimed, became the biggest man-made forest in the world. The area became more productive under the trees than it would otherwise have been.

Sheep graze between the long racks of flax leaves which have been stripped from the main plant and hung out to bleach and dry in the sun.

The failure of the flax industry

Products based on *Phormium tenax*, the native plant also known as New Zealand flax, became a significant export in the nineteenth and early twentieth centuries but the industry never fulfilled its promise because it was extractive rather than based on cropping. The plant and its many varieties grew abundantly in swamps – so abundantly and so accessibly that Maori hadn't needed to plant it as a crop, while using it widely for clothing and cordage.

When export demand first built up for the product in the nineteenth century, the processing expertise of Maori women was used. As exports increased, mechanisation was needed to lift production. Export volumes and receipts reached 150 tons, worth £4674, as early as 1855 but then varied from year to year, reaching 22,653 tons worth £595,684 (three times as much as cheese earned) in 1903.

Once mechanised, flaxmillers, operating under freehold or licensed rights, set up their plants near swamps, harvested and milled the various strains of the plant, indifferent to the various qualities of each – such as variations in strength, length and colour. They operated much as a sawmiller did in stands of native forest. Little if any attention was given to choosing the superior varieties and then breeding and cultivating them as was done with food plant crops.

As competition developed from Manila hemp and Mexican sisal, an attempt was made in the 1930s, through the Council of Scientific and Industrial Research, to support hybridisation research and the development of high-quality strains. But it was too late. The Admiralty in Britain used phormium for some marine cordage and the industry scored a large contract to supply the United States Navy with rope early in World War Two when production was averaging only 6000 tons a year. By the 1950s, annual production was at 5000 tons as it continued its historic use in woolpacks for the New Zealand wool industry. However, phormium production faded soon afterwards in the face of the strength and flexibility of artificial fibres.

forests planted in the twenties. The production of pulp and paper from the central North Island forest had been talked about since the earliest years of reafforestation but the capital required was huge, and the industry cyclical and internationally competitive. One investigation into paper production in the early years of the twentieth century took very little time to dispense with the idea because the amount of capital needed was prohibitive at that time. However, in the early 1950s the town of Kawerau in the Bay of Plenty was set up to serve the establishment of a nearby plant for pulp and newsprint production. Thus, forest products of many types supported a major industry in the immediate post-war period. The foresight of a couple of generations of governments and foresters had provided the country with an asset of enormous value.

Farm forestry schemes subsidised by the government spread the area of forests, and the replanting of trees in hill and high country helped cut the damage by erosion which seriously affected regions like Poverty Bay in the post-war years. Less than one-quarter of more than six million hectares of indigenous forest – mainly beech, kauri, rimu, tawa and taraire – remained in private hands, with the

The Tasman Pulp and Paper Co Ltd was formed in 1952 and pulp and paper mills were constructed in the Bay of Plenty (foreground). To supply labour for the huge industrial plant, the satellite town of Kawerau was built nearby (background). This picture was taken at the end of 1955.

In 1955 a huge paper machine at the Tasman Pulp and Paper mill in Kawerau was nearly ready to start producing newsprint from raw wood pulp twenty-four hours a day, seven days a week. The mill was the first in New Zealand to make newsprint paper on a large scale.

rest owned by the government and managed by the Department of Conservation. Legal provisions control how private owners harvest and sustainably manage their forests.

During the last half of the 1970s private plantings of exotics were outstripping those by the state for the first time. Many farmers combined livestock farming with forestry, taking advantage of government incentives to do so. By the 1980s sheep and beef cattle farms had over 100,000 hectares in exotic trees. The timber mills were sited within the region of the forests, which were progressively milled and replanted in a sustainable circuit. The harvesting and processing of timber required such large amounts of capital that two major industrial corporations gained control of more than a quarter of the forests.

Plantation forests covered about 1,830,000 hectares as plantings continued in the twenty-first century, with more than 100,000 additional

A well-cared-for plantation of pines near Taupo, some of the millions of radiata pine trees that clothe the landscape of the central North Island.

hectares in 2003, followed by a fairly steep decline. The government remained involved. Recent plantings included the East Coast Forestry Project set up by the government in 1992, originally to plant 200,000 hectares of commercial forest. The project was severely modified but about 40,000 hectares were planted on land at risk from erosion.

Government grants were available to assist landowners with the establishment and management of the forests. The Crown Forestry Group, on behalf of the government, managed twenty-five exotic forests, sixteen of them leased from Maori landowners.

Total exports amounted to more than $3.2 billion. Some progress was being made towards the ideal of adding value by higher levels of processing, with exports of sawn timber and finished products, mainly to Australia and the United States, increasing in value. But the sale of logs to Korea, Japan and China remained the biggest single end use.

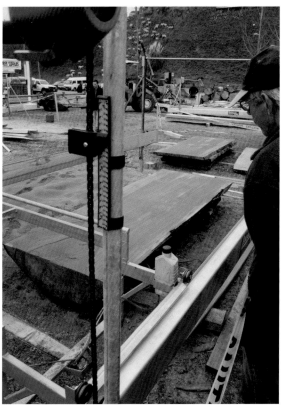

The central North Island was where the big plantations of indigenous trees were first grown, harvested and milled and it remains the centre of the industry, although early in the twenty-first century the first drop in the country's planted forest area in decades occurred and most of the land not replanted was in this area. Here, rimu timber is being milled for use in the manufacture of furniture.

CHAPTER NINE

NEW FRUIT FOR OLD

No doubt existed from the time the first tree sprung its crop that temperate-zone fruits would thrive in New Zealand. Missionaries brought the seed in and Maori, with no experience of trees that bore so abundantly, took them around the North Island in the manner of the mythical Johnny Appleseed in the United States. Wherever they went they sowed the seeds.

The director of the Department of Agriculture's horticulture division, J. A. Campbell, wrote in the mid-1930s, 'before organised settlement took place and for many years subsequently, groves of stone fruit trees, particularly peaches produced from stones planted by the Maoris, were found bordering practically every watercourse in the North Island. These trees bore heavy crops of beautiful fruit, and continued to do so until age and the ravages of introduced diseases eventually wiped them out.'

The traditional temperate-zone stone and pip fruits were fundamental to the domestic industry here for a century; and over most of that time apples and pears dominated a steadily built-up export trade – until the kiwifruit and grape bonanza of the 1970s and 1980s which brought about a revolution in New Zealand agriculture unmatched since refrigerated shipping. Kiwifruit and grape vines, carried on wooden frames in long, straight green lines, redecorated countryside once dotted with sheep and cows. Both did well in a variety of regions but kiwifruit grew more plentifully in the Bay of Plenty than elsewhere and grapes thrived especially in Marlborough and the sun-dried east coast of the North Island. Kiwifruit originated in China but were reincarnated into an international delicacy in New Zealand. Grapes are as old as civilisation in the Middle East and Europe but some varieties found a new, reinvigorating home here for the making of wine.

In the early days of settlement, the stone and pip fruits were the staples and, with the concept of biosecurity unknown in its modern sense, immigrants continued to bring from other countries plant material on which diseases could easily survive the long sea journey. Codlin moth, woolly aphis, scale pests, apple and pear scab and a range of fungal diseases were soon rife, and it wasn't until the Orchard and Garden Diseases Act 1903 that action was taken to curb the unmitigated spread. Old and neglected orchards were destroyed and a ban placed on the sale or distribution of infected fruit or trees. Two years later, arsenate of lead became a standard and effective control for codlin moth.

The earliest orchards in the sense of organised planting of fruit with the intention of harvesting for sale were sited on Auckland's North Shore in the 1850s. Orchards tended to be planted close to concentrated settlements of people in towns and cities because of the perishable nature of most ripening fruit before chilling facilities were available. Unlike today, peaches, nectarines, apricots, apples, pears, plums and cherries had their brief seasons of high availability in summer and autumn, during which consumers gorged themselves while they could and then waited in anticipation of

Grapevines have redecorated a landscape once dotted with sheep in the north-east of the South Island, as in this vineyard near Blenheim. Their frames in long straight green lines reveal the design of a new diversified New Zealand agriculture.

fructuous times the following year. The same pattern applied to tropical imported fruit. Generations of New Zealand children received an orange in their Christmas stocking – a rare and exotic gift. Before World War Two, lemons were grown on nearly 160 hectares in New Zealand and oranges on 405; by the end of the century, mandarins, oranges, lemons, tangelos and grapefruit were grown on more than 2000 hectares. Most of these fruits are subtropicals and do well here only in the mildest regions of New Zealand's temperate climate: Northland, Auckland, Bay of Plenty and the East Coast. Exports from Australia and the United States continued to bridge seasonal gaps.

The first apples and pears, like so many plants and animals, arrived with Samuel Marsden, who planted trees at the Church Missionary Society's Kerikeri base in the Bay of Islands. Twenty-five years later, the society reported that the pip fruit trees were producing abundantly at most of their stations.

The development of an export trade was inhibited for many years by the perishability of fruit, except that apples and pears survived better than the soft stone fruits. Orchardists sent the first export consignments of apples to Chile in 1888 and Britain four years later. From the first decade of the twentieth century, a trade built up with South America which took 68,000 cases of apples in 1914 before World War One brought it to a temporary halt. By this time, the main apple and pear production was moving inexorably towards Hawke's Bay and Nelson – still the principal source of supply for both the domestic and export markets over a hundred years later. The climates in the two regions played the biggest part in this shift. Apple growers had discovered they could make a living from land that was marginal for pastoral farming – often land that had suffered severely from the depredations of rabbits – as long as the weather was right. Peaches found their most salubrious home in Hawke's Bay, and apricots in Central Otago.

Over the years, apple and pear orchardists followed generally in the footsteps of other primary producers towards co-operative distribution and marketing, helped by the reigning governments. A

Pears have always been a lesser product than apples but still a significant part of the export market.

Opposite: Kerikeri has long been the area in which citrus fruits have thrived because it has the mildest of New Zealand's temperate climate. Pictured are pickers of Valencia oranges in a Kerikeri orchard in 1949.

The New Zealand Government for more than a century endeavoured to ensure that New Zealand export products were of a standard that would keep the reputation of primary industries high. Pictured are government fruit inspectors in 1939 examining apples destined for export (individually wrapped in tissue paper).

Fruit Control Board was put in place in 1924. It controlled all exports under one label, and four years later shipped a million cases.

Fruit and vegetable growers have always faced a more precarious existence than those involved in traditional New Zealand agricultural pursuits. Contrary weather – frosts, hail, heavy winds, floods – disease and bird strike could eliminate crops, or spoil them for export where high quality standards had to be met. While some technological developments helped ameliorate frost and bird damage, fruit-growing remained a risky business. Equally, overproduction in good seasons could corrupt markets – especially when it matched gluts from overseas producers.

On occasion over the years, the New Zealand industry reacted to economic forces to reform itself. In the decade after World War One, orchards got bigger and fewer, with about 7500 growers on an estimated 50,000 acres shrinking to fewer than 6000 on about 27,000 acres. And again in the 1950s, the industry was persuaded to cut back the number of apple varieties from about 140 to fifteen, and pear varieties from more than forty to nine. In more recent years, the number of varieties of apples grew again, although some varieties faded in competition with such desirable new commercial brands as Gala, Royal Gala, Fuji and Braeburn, which make up more than seventy-five per cent of exports and on which international promotion is focused. Jazz apples entered the market in the new century. Except for the Japanese variety, Fuji, the new apples were the result of research and development in New Zealand where scientists were always striving for a better apple and pear.

During the early years of World War Two, apples and pears were crowded out of the limited shipping space between New Zealand and Britain and the crop was sold domestically. The government subsidised growers when the price slumped below the cost of production. Apples were sold cheaply – even given free to schoolchildren – and, according to one account, consumption reached 265 apples a year for every New Zealander. Substandard fruit was used as pig fodder.

The government in 1948 set up the Apple and Pear Board in much the same shape as the other producer boards, with grower and government representation. Subsidies to growers were removed and the board operated a guaranteed price to level out growers' returns between fluctuating prices over the seasons – a mechanism that lasted until the 1980s. In 1952 the board was given the power to control all purchasing and marketing.

The year the board began exporting to the United States, 1959, international production was increasing and squeezing prices, and also forcing up the standard of quality required. The move into the United States was a bid to widen the product destination which had for decades been predominantly Britain. At this time, too, the growers began using new techniques, successfully experimenting with pruning methods and smaller trees – some with central leaders in pyramid shape, some as espaliers on wire frames. Combined with more effective disease and pest control, these developments gradually doubled the yield per hectare. More efficient storage and distribution systems kept apples on the domestic market virtually all year round.

As the 1960s moved into the 1970s, the board launched a major marketing drive to diversify away from Britain and

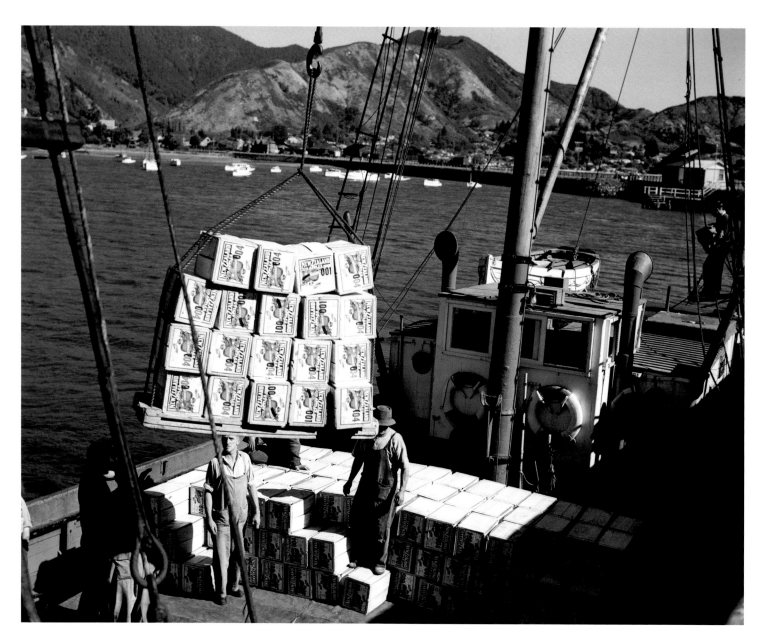

Apples are loaded on a scow in Nelson in 1948, one of the two traditional apple-growing regions (with Hawke's Bay), for Wellington, where they will be transhipped to vessels to take them to London.

the five or six other established markets. Eventually New Zealand apples went on sale in more than forty international markets. From the 1960s, as the demand for higher quality rose, surplus apples that didn't make the export cut were used to develop special consumer products such as apple juice (combined also with other fruit juices) and bakery trade items such as apple pies and apple slices. This ancillary processing industry also helped absorb excess production in years when international markets were jammed with apples. Products included a range of apple, pear, carrot and berryfruit juice concentrates. Through an Australian-based subsidiary, Fruitmark, the New Zealand industry also sold concentrates, dried fruits, fruit pastes and dehydrated fruit and vegetable products.

In the 1980s the board ended its guaranteed price scheme and adopted a bold strategy similar to that of the Dairy Board, buying as well as selling fruit from around the world, trying to command the trade as a kind of international broker. Meanwhile, exports reached more than seven million cases in 1984 and doubled to fourteen million cases only seven years later. This expansion of sales occurred without the ability to export to Australia since 1920, when a ban was invoked on the grounds that imports would introduce fireblight, a tree disease endemic in New Zealand. New Zealand claimed widespread scientific support for its case that fireblight could not be carried on picked fruit. The World Trade Organization agreed and Japan, which had also imposed a ban on New Zealand apples, decided to comply with the WTO ruling. However, for Australian apple growers immunity from competition was powerful motivation for keeping pressure on their government not to relent on the ban even after eighty-five years. The issue remained unresolved into the twenty-first century, still causing trans-Tasman friction.

In 1991 what was by now the Apple and Pear Marketing Board decided to brand its apple and pear exports under Enza, a name chosen, said the board, because it was short and punchy and clearly evoked New Zealand. International competition was heating up at the time, and became increasingly intense as production grew, especially in the southern hemisphere, which posed a seasonal challenge to New Zealand. In difficult times, dissension grew among growers, despite the fact that the board was owned and governed by them as a co-operative. Enza had expanded too far as a bureaucracy,

Apple orchard in Richmond near Nelson. The area bearing apple trees declined as overproduction swamped the European market, including southern hemisphere fruit, and left growers with deep financial problems, but the efficiency of the industry in New Zealand had always been admired.

Opposite: Kiwifruit growing strongly in Te Puke, the place where the fruit got its real commercial start and where it is still a dominant commercial activity.

Te Puke orchardist Jim MacLoughlin was the man who invested his faith and money in kiwifruit, confident the sweet-fleshed, good keepers would become an international commodity – and he was right.

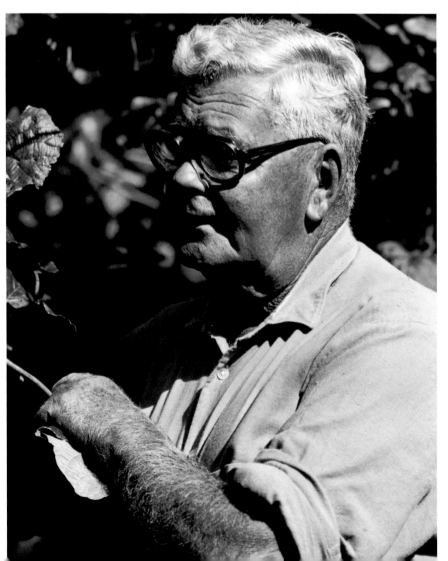

according to many, who insisted it was costing them too much money, even though it funded a lot of research, including work on new varieties. Many companies were trying to export apples and they applied pressure against the single buyer-seller concept and contributed towards the imminent demise of the board.

The board's control ended and the industry corporatised with the passing of the Apple and Pear Restructuring Act at the end of 1999. The Act turned the board into a limited liability company, to be known as Enza Ltd. Shares were distributed among growers in proportion to supply levels over a period of years. Enza became just one of a number of selling and distribution players. The industry entered a consequent period of disarray as it tried to settle into the new structure – or lack of structure to be more precise – while export companies competed against each other in world markets.

At a time of an international glut of apples and with kiwifruit and grapes enjoying steadier growth, the area of land in apples declined. Trees were still being pulled out in 2005 as apples continued to concede land to grapes in Nelson and on the east coast of the North Island, and the area in apples had already dropped from 14,541 hectares in 1999 to 11,715 in 2003. The troubles came to a head in 2005 when sales in a Europe overloaded with apples slumped and so did the price, leaving growers and packers in New Zealand with deep financial problems. Faced with a calamity, twenty-nine growers and exporters agreed to form a market panel to administer a New Zealand pip fruit brand, complete with logo, using a quality imprint, or trustmark. The task that lay ahead was to convince all those involved, especially exporters (and there were more than a hundred of them), that demonstrable benefits would be gained from joining the panel and accepting its constraints along with its marketing plans.

The efficiency of the New Zealand industry had long been admired, and although the country was distant from markets it still had advantages: yield of apples per hectare was matched elsewhere in the world only by the Netherlands, and the scientists were leaders in the business of developing new consumer-attractive varieties, such as Jazz. Other countries would have been similarly hurt by the glut and would, perhaps, pull back on production. So the belief remained that, in the longer term, apples would retain a place, albeit diminished somewhat, in the New Zealand economy.

The kiwifruit story buzzed with glamour when it broke at the end of the 1970s, enhanced by the fact that the growers who had risked all on the future of the brown furry fruit became millionaires in a fairly short time on relatively small areas, most famously on No. 3 Road, Te Puke. Within a few years, other growers in other places followed suit and the fruit supplanted apples and pears as the biggest horticultural export earner; so an orchardist called Jim MacLoughlin and those who followed his entrepreneurial example did much for the New Zealand economy as well as for themselves.

The kiwifruit plant, a Chinese native, had been transported around the world by curious travellers and botanists since the nineteenth century, but it took New Zealand orchardists to recognise its broad appeal and commercialise it. The fruit was then called Chinese gooseberry because of a vague similarity in appearance to the gooseberry when cut

open, not from any similarity in outer appearance or flavour. The plant held a place in the Royal Botanic Gardens, Kew, from the mid-nineteenth century, and early in the twentieth a British nursery firm, James Veitch and Sons Ltd, listed *Actinidia chinensis* under 'Novelties for sale' but apparently sold only male plants. All the buyers got was a vine with nice leaves and flowers.

In 1904, as the plant was being disseminated around the world – often by missionaries and more as a curiosity than a commercial prospect – Miss Isabel Fraser, principal of Wanganui Girls' College, arrived home from visiting her evangelist sister in the Yangtze Valley, in China. She brought with her some Chinese gooseberry seeds. They were propagated by a Wanganui horticulturist and by 1920 a number of nurserymen were growing and selling the plants, among them a man called Hayward Wright. He developed the Hayward cultivar, which eventually became the variety on which the export industry came to rely.

Chinese gooseberries became popular on the domestic market as orchards in Wanganui, Northland, Auckland and the Bay of Plenty began producing reasonable volumes of the fruit. By 1948 demand was small compared with tree tomatoes, or tamarillos. Indeed, as late as 1965, 184 acres were under kiwifruit and 377 under tree tomatoes. However, Te Puke orchardist Jim MacLoughlin invested his faith and money in his belief that kiwifruit, sweet-fleshed and good keepers, had the potential to become internationally desired. He made his first commercial planting in 1934, had seven acres planted in vines by 1940 and forty-five acres by 1955. He pushed to export the fruit, teaming up with another Te Puke grower. Together they shipped out 2000 cases in 1952 and the trade grew from there. MacLoughlin was later tagged 'Father of the Kiwifruit Industry'.

However, Chinese gooseberry was never considered a name that would resonate with consumers and was especially unappealing in the United States because China was a communist country. The first alternative name suggested was 'melonette' but a copywriter at Catts Patterson Advertising Agency, Chris Cole – later a writer, journalist and publisher – came up with a list of thirty possible names for the agency's client, exporters Turners and Growers.

It included 'love berries', 'fuzzyberries' and 'kiwiberries', the last of which appealed to Jack Turner. He consulted a botanist who told him they were fruit, not berries. So kiwifruit they became, in 1959, and the name stuck, becoming a generic name all over the world. Turner later lamented the fact that the name was not protected, but doubt prevailed whether a combination of two such common words could have been registered anyway. By the 1970s New Zealand growers couldn't keep up with the demand and horticulturists in other countries welcomed the chance to buy the highly successful Hayward plants to compete. Orders flowed in to New Zealand nurseries from Australia, Italy, South America, South Africa, Israel and other countries. One prominent plant nursery firm accepted an order to sell to a French agent 10,000 Hayward plants a year for ten years. An argument raged over whether the country was wise to sell the plants. Some claimed other climates would make it hard to achieve top quality; that it would be years before foreigners would be able to learn the idiosyncrasies of the plant; and that even when they did New Zealand would still enjoy a seasonal advantage. Others, less sanguine – and they turned out to be correct – claimed foreign production would expand rapidly, compete in European markets with the New Zealand product, and adversely affect prices. Forty years after the plants were sold, Italy grew more kiwifruit than New Zealand, although New Zealand still controlled much of the international trade.

The number of hectares in kiwifruit in New Zealand grew from fewer than 10,000 in 1982 to just on 18,000 three years later and peaked at nearly 19,000 in 1988 before beginning to decline. Production per hectare, however, steadily increased and by 2004 the area sown was below 11,000 hectares again. In the 1970s and 1980s pioneer growers like Jim MacLoughlin were generously rewarded with profits for their foresight, their dogged persistence and their flair. One million trays were exported in 1976, and only two years later more than two million trays were shipped out, as well as several hundred tonnes of processed fruit (mainly canned).

New Zealand experienced a kiwifruit frenzy in the boom years of the 1980s as farmers converted

Lasting affection – a media affair

A residual love of the land lies in the hearts of most New Zealanders even though they have been severed from any close relationship with it for at least a generation. What else could explain the continued existence of *Country Calendar*, this country's longest-running television programme? For decades its familiar introductory theme music on a Sunday night drew many thousands of farmers as well as townies to their television sets.

The programme – with the pronounced objective of 'informing viewers about agriculture and the primary industries' – first flickered onto our television screens in glorious black and white in 1966. Its first presenter, Fred Barnes (above), wrote later that he and his staff interpreted the objective in more detail as 'communicating useful information to farmers; creating an awareness of new agricultural developments and techniques; encouraging sound farm management practices; and presenting the rural scene in a way which would interest all sections of the community'.

Barnes, like most of the original *Country Calendar* staff, was recruited from radio, which itself had for many years served farmers with excellent information programmes. Among the earliest of those who converted to television was Bruce Broadhead. A Lincoln College graduate, he was crucial in transforming farming news from radio to the television format. *Country Calendar* was based to some degree on the style of his 1YA radio show, *Country Journal*. But the broadcaster who had the longest association with the television programme was Frank Torley, an experienced agricultural journalist who never lost his flair.

Country Calendar was later moved to Saturday nights and, after a number of sponsors over the years, became a flagship of the National Bank of New Zealand. The programme also became famous for its richly funny end-of-season spoofs.

Farmers were well served by print media right from the nineteenth century. Newspapers and magazines, many with large circulations penetrating deeply into their communities, catered for rural families, providing space for gardening notes, women's sections, children's contributions and local, Australian and English fiction, as well as reports of agricultural and pastoral society shows, and cattle and sheep breed notes. During the first half of the twentieth century the rural magazines were also a valued source of income for notable New Zealand writers such as Robin Hyde and James Cowan.

Agricultural specialists were among the best and the best-known journalists in the country. Among these skilled and influential craftsmen were Peter Freeth of the *New Zealand Herald*, Doug Fyfe of the *Christchurch Press*, Tim Tyrer of the *Dairy Exporter*, Ronald Vine and John Cornwell of the *New Zealand Farmer*, and Gilbert (Gib) Neale of the *Journal of Agriculture*. The print media thrived unchallenged until the 1930s when its information services were complemented by radio programmes dedicated to rural life.

At the end of the 1950s David Yerex returned from Britain, where he had been a member of the guild of agricultural journalists, and advocated the formation of a similar guild in New Zealand. Yerex, a great all-rounder who wrote in several genres, later moved easily between public relations and journalism. He persuaded Tyrer to become the interim president and the New Zealand Guild of Journalists was formed in 1959. It grew to be a significant media organisation by the twenty-first century, with 120 members representing those involved in the full span of print and electronic media as well as public relations.

land to grow the fruit and city investors found a way to make profits and avoid tax. Gradually the industry groped its way to sensible joint methods of processing, distribution and marketing, and, consequently, reasonable profits were achieved in most years. But when the crop jumped from twenty-nine million trays in 1986 to forty-six million the following year and multiple agents were selling, the market staggered. Some growers thought the end of the golden days was in sight and got out of the industry but most began to think of a single marketing entity.

That happened in 2000 when Zespri Group Ltd was formed, a single export entity with an integrated supply and delivery system, wholly owned by growers. Kiwifruit would henceforth be marketed under the Zespri brand – Zespri Green and Zespri Gold – causing some commentators to joke that a kiwifruit is not a kiwifruit only when it is produced by Kiwis. Zespri Group Ltd immediately became the world's biggest kiwifruit marketer. By 2004 export sales were approaching $700 million, representing almost thirty-five per cent of New Zealand horticultural exports – well ahead of apples and pears.

Zespri Gold – a soft, yellow-fleshed variety – was launched commercially in 2000. It had been developed over eleven years at the Te Puke Research Orchard and this time plant variety rights were taken out internationally. Trial export consignments had been shipped in 1997 and seven years later it was achieving enough market success to encourage many Zespri Green growers to make a shift to gold. The fruit was already widely grown in New Zealand and, under licence, in a number of foreign countries, including Italy, France and Chile.

Kiwifruit and apple exports were worth more than fifteen times as much as all other fruits together, and faced no immediate challengers to their pre-eminence. Avocados, originally from South America and worth around $30 million a year in export returns in 2005 (mainly from Australia), were first grown commercially in New Zealand in the 1930s. Production increased by eighty-five percent in the year to 2005, predominantly because of a rapid expansion of plantings over the previous five years. A tropical fruit, avocados need care and attention in this temperate climate, being sensitive to frost and wind and preferring free-draining soils. They occupied 550 hectares in commercial orchards. The highest per capita consumption in the world is Mexico (more than eight kilograms) where it is the main ingredient of a national dish, guacamole, but consumption increased rapidly in the 1990s in the United States with its growing population of Latinos.

Two other fruits of South American origin, tamarillos and feijoas, were relished locally but were insignificant export earners, depending almost entirely on the United States market. Tamarillo trees are soft and need complete shelter from wind, while feijoas are delicate fruit and susceptible to frost. High hopes were held for tamarillos around the time kiwifruit were making an impact but export returns have never reached $1 million a year. They carried the name 'tree tomato' from the time they arrived in New Zealand in the nineteenth century until 'tamarillo' was coined as a more marketable alternative by Mr W. Thompson, a member of the New Zealand Tree Tomato Promotions Council, in 1967. He joined the Maori, 'tama', which he translated as implying 'leadership', with what he thought was a Spanish ending. New Zealanders seem to have a gift for nomenclature because the name took and is now used in most parts of the world. Tamarillos are widely grown in gardens in many other countries but they have never achieved mass appeal. Commercial crops are not numerous, perhaps because the fruit has a tart rather than a sweet flavour.

Persimmons have been grown commercially in New Zealand only since the 1970s, but production built up strongly during the 1990s as the fruit proved profitable and in export demand (worth around $7 million a year), mainly from Asia, where the plant came from originally.

Grapes grown for wine seldom contest for land with the subtropical fruits. Like apples, grapes were first introduced by Samuel Marsden when he arrived here with a variety of animals and temperate-zone plants to provide food for missionaries and Maori. James Busby, the New Zealand Resident who was the first local vestige of British government, arrived in 1833 and planted vines at Waitangi. He had studied grape-growing and wine-making in France and written a book on the

Grapes had been around in New Zealand for more than 150 years when the wine revolution broke out in the 1960s. These women are picking grapes in West Auckland, mainly for eating or for sweet table wines and sherries, which was what the market still wanted.

subject. When he was in New South Wales before coming to New Zealand he bought land in the Hunter Valley, now a famous wine-making area, with the intention of growing grapes and producing his own wine.

When he was appointed Resident, Busby was in Britain, and before returning to the colonies he visited France and Spain, collecting vine cuttings which he distributed throughout New South Wales, South Australia and New Zealand. By 1838 Busby had extensive vegetable gardens and a vineyard and was making wine which he sold to British troops stationed in the Bay of Islands. Within a few years, French settlers at Akaroa had their own vineyard, and French Catholic missionaries began wine-making in Hawke's Bay in 1865 at a place called Meanee, still a wine-making centre 140 years later.

A viticultural centre was set up by the government at the beginning of the twentieth century at Te Kauwhata to import and investigate grape varieties and experimentally make wines. But despite bouts of enthusiasm for New Zealand as a potentially world-class site for growing grapes and making quality wine, no momentum gathered for the birth of an industry. One reason may have been the British tradition of drinking beer, which many of the puritan population thought created enough social problems as it was. Few New Zealanders had acquired a taste for wine. The prohibition lobby was a powerful force within government and would certainly have discouraged any prospective investors in a wine industry. One pocket of production, though, was in the north and west of Auckland among Dalmatian and Lebanese immigrants who brought with them a tradition of wine-making, and one that survived the twentieth century.

The government viticulturist and oenologist for many years Francis Berryman wrote in the 1960s

The first commercial grapevines in Marlborough were planted in 1973 and thirty years later the province was home to a booming wine industry. A large harvester runs along the vines picking the grapes which are then conveyed to an adjacent bin.

(about the time cheese-makers were daring to try camembert at Opouriao): 'Although the New Zealand climate is recognised as being particularly well suited to the production of table wines, the people, like most who are of British descent, prefer the stronger sweet wines, such as sweet sherry, port, Muscat and Madeira types.' He was right, as any New Zealander of that time would remember. Sherries and ports were still being consumed in large quantities and the national attitude towards table wines as the draught of effete Europeans was only just beginning to fade.

Then, astonishingly, within the space of a mere twenty years, table wines and Continental cuisine became all the rage and wine-makers of skill and taste were producing white wines acceptable to sophisticated drinkers anywhere in the world. Berryman had written in the 1960s: 'Up to the present the high cost of grape production . . . and the traditional bias of overseas wine markets have discouraged concerted efforts to develop a wine export trade.' That situation also changed dramatically.

In Berryman's time eighty-five per cent of the vineyards were in Auckland (425 acres) and Hawke's Bay (387 acres) and that was the way it stayed for some time. An earlier oenologist, Romeo Bragato, had found that the European *Vitis vinifera* varieties of grape predominated in Hawke's Bay but that in Auckland Franco-American hybrids, including *Vitis labrusca*, were most common because they did better in higher rainfall regions and were largely resistant to phylloxera which was present, and deadly, in *V. vinifera*. It was some years before the lower yield but higher quality (for wine) European vines again became important, mostly grafted onto phylloxera-resistant rootstock.

The Department of Agriculture knew from the end of the nineteenth century that the best places to grow wine were Central Otago and the north of the South Island as well as Hawke's Bay. In 1895 Bragato, born in Yugoslavia and trained in Italy and France, came to New Zealand as a consultant to the government and reported that Central Otago was an ideal site for growing grapes for wine. 'There's no better country on the face of the earth for the production of Burgundy grapes,' he wrote. He also suggested that Canterbury, Nelson, the Wairarapa, Hawke's Bay and some places in Northland had excellent potential. Although he did not visit Marlborough that time, he was convinced of the value of the region at the top of the South Island. Bragato returned to New Zealand in 1902 to set up the experimental station at Te Kauwhata for the Department of Agriculture but left in 1909, disillusioned by lack of government support and the general indifference of New Zealanders to wine. Not only was the department focus firmly on the pastoral industries, government advisers had decided to leave wine-making within the regions in which it already existed and not to spread it thinly, as one of them described it, around the country. It took private interests to sniff out the best vineyard sites much later in the century.

The powerhouse of the industry by 2005 was Marlborough. The first vines were planted there in 1973, but thirty years later fruit trees were being pulled out and sheep shooed off the grass so land could be put down in grapes. Just under 10,000 hectares were in grapes by 2005. The area had grown by more than a thousand hectares a year since the new millennium, and was five times larger than in 1996. Alluvial loams over gravelly subsoils in the Awatere and Wairau valleys, the longest annual sunshine hours in the country, cool nights, and a long growing season all contributed to the success of the region. The big grape there was sauvignon blanc, followed by chardonnay, pinot noir and riesling. Neighbouring Nelson's area in grapes had grown from ninety-seven hectares to nearly six hundred. Over the same period, Central Otago had been developed and was producing delightful vintages, including pinot noir, and some new niches had been found to confirm the assessment of Bragato a hundred years earlier. Southern Wairarapa began to produce lovely pinot noir as well. The area planted there doubled in three years to 758 hectares. The whole country was in the grip of the grape.

Some primary production traditionalists, knowing the history of booms and busts, became concerned at the rate of investment of land and money in grapes but wine export volumes and returns in the largest markets held up well over ten years. Britain and Ireland bought most, followed by the United States, Australia and Germany.

Berryfruit also depend on local demand but blueberries, strawberries and boysenberries did have counter-seasonal export markets that may vary from year to year, depending on the effect on prices of local production in those markets. Overall export returns have reached $14 million. Blueberries rode high in the new century on the back of health claims but, along with boysenberries and strawberries, faced disease concerns. Blackcurrants went into broad acre production, particularly in Canterbury, and occupied a greater planted area (1300 hectares) than all other berryfruit combined, supplying processors and manufacturers of some of the most popular fruit drinks on the market.

In 1960 no one could have foreseen the transformation of the fruit-growing industry in New Zealand. Although export earnings from pip fruit, kiwifruit and wine were still dwarfed by dairy and lamb returns, the explosion of horticulture gave some much needed breadth and balance to the country's export profile.

Boysenberries (pictured) and other berryfruits find counter-seasonal opportunities overseas, which vary from year to year, building export revenue of up to $14 million.

CHAPTER TEN

THE VEGETABLE BUSINESS

James Wattie was a Hastings boy who became a pioneer of the fruit-canning business and formed J. Wattie Canneries Ltd.

The canning of peaches by Watties helped end the era when many housewives did their own bottling when fruit was available and cheap at the height of the season.

THE FRUIT AND VEGETABLE INDUSTRIES were transformed by a series of changes after World War Two: canning; advances in chilling and freezing and in packaging materials; the advent of supermarkets; air freight and faster transport on better roads; and the demise of the home vegetable garden.

The keeping qualities of some varieties – pip fruits, kiwifruit, onions, squash and potatoes among them – ensured an export market even in the days of relatively primitive packaging and shipping, but stone fruits and many vegetables had to wait until processing could give pause to their rapid, natural deterioration.

James Wattie was a Hastings boy who studied accountancy as he worked in companies associated with the fruit, vegetable, arable and meat industries. Canning was in no way new to New Zealand but Wattie saw prospects for preserving Hawke's Bay's prolific ability to grow stone fruits, particularly peaches, and vegetables such as tomatoes, peas, beans and corn. He was a pioneer, forming J. Wattie Canneries in 1934, setting up a factory in Hastings and gradually building up the country's largest canning business. As World War Two ended, the company was building up to produce a full range of canned and bottled foods: tomato and other soups, canned fruits and vegetables, jams, and the famous tomato sauce which became one of the most ubiquitous items in our grocery shops and dairies.

Although Watties – as the company became universally known – bought a farm in Hawke's Bay in 1945, it was soon processing enough fruit and veg-

etables to encourage market gardeners and some farmers on the fertile Heretaunga Plains around Hastings and on the Gisborne flats to move into cropping on a larger scale than ever before. Apart from a wide variety of vegetables, apricots from Central Otago and berryfruits from other regions soon found a market for canning and freezing. In 1951 the company established a factory in

Gisborne after persuading farmers there to plant sweet corn, and more than four million cans of corn came off the assembly line in the first season. In 1970 a factory was set up in Christchurch, mainly to quick-freeze peas.

Watties was not the only company that took up canning and freezing food during the post-war period but it was the biggest and best known until it was bought out by the international food-processing giant H. J. Heinz Co in 1992. Lever Brothers also began canning in Hastings and in Motueka in the 1950s and became a big producer of frozen and dehydrated foods. It was also an early producer of frozen meals, which followed quickly thereafter and became especially big business with the rise of supermarkets.

As trade became more and more global, growers everywhere had to compete on price, especially orchardists. For example, canning costs had long been relatively high in New Zealand because of wage rates and small domestic runs. Growers supplying large processing companies in more populous countries could achieve such economies of scale that Hawke's Bay peaches and Central Otago apricots faced competition from South American and South African fruit. In those regions margins may have been thin but they ran across huge volumes.

Before the 1950s vegetable growing was largely confined to local market gardens selling through greengrocers, and to home gardens. The bottling of fruit and making of jam were chores done annually

The range of canned and bottled foods expanded as the twentieth century progressed. Pictured is fruit being boiled to make jam for bottling by Thompson and Hill in Auckland.

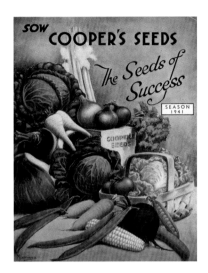

by many housewives, but all this home processing changed as cities grew bigger and New Zealanders rapidly became urbanised.

The refrigerator with a freezer compartment was a presence in every household during the two decades after the war and, by 1955, Watties was quick-freezing peas in both Hastings and Gisborne. The vegetable industry remained focused on the domestic market – as it is today – but counter-seasonal advantages allowed New Zealand growers to fill niches in other countries, especially the northern hemisphere, via fast and efficient transport systems, including containerised shipping. High-quality, long-keeping fresh vegetables such as onions, squash and potatoes became significant export earners, filling out-of-season gaps in local supplies in a number of countries. In the 1980s consumers in all developed countries abandoned the idea that seasons should be allowed to affect the supply of fresh produce, so New Zealand growers and pack houses began to export hydroponic lettuces, bagged salads, packaged greenhouse capsicums and tomatoes, and other lightly processed ready-to-eat products, as well, of course, as frozen, dehydrated and canned vegetables.

By 2004 exports of fresh vegetables to dozens of countries totalled nearly $220 million, with onions ($92 million) and squash ($53.5 million) the big winners, but with capsicums ($24 million), carrots ($10 million), tomatoes ($7 million and rising), and asparagus ($7 million) emerging with potential. New Zealand became the largest exporter of fresh onions to the European Community, and of squash (kabocha) to Japan. Processed vegetables earned $265 million, mainly from beans ($40.3 million), potatoes ($66 million), peas ($39 million), sweet

National Agricultural Fieldays

The National Agricultural Fieldays are held in June every year at Mystery Creek, near Hamilton, and after thirty-six years the organisers were calling it the biggest annual event in New Zealand and one of the biggest for farmers anywhere. The claim was based on an attendance in 2005 of comfortably more than 120,000, total orders and sales worth close to $150 million over the three days, and overall orders and follow-up sales of nearly $300 million. Not bad for a city that itself had a population of only around 120,000. One advantage of Hamilton as host city was its centrality to perhaps the finest pastoral farming land in the world.

This New Zealand festival of farming grew from the union of two earlier annual events. The marriage broker was a prominent Waikato farmer, John Kneebone, who wrote a letter to the farming editor of the *Waikato Times* in the 1960s, suggesting that:

> [O]ur famous Ruakura Farmers' Week and our Waikato A&P Winter Show coincide so that New Zealand farmers could have their annual outing in Hamilton . . . British farmers take their holiday to go to London to see the machinery [and] the carcase meat exhibits [at the Smithfield Show] . . . As well as a well earned holiday, New Zealand farmers from near and far could see Ruakura at its best, the winter show, the machinery, and many of the attractions of the Waikato.

A group of influential people with wide-ranging interests followed up this idea and the first National Agricultural Field Days (the spelling was

later made 'distinctive' as Fieldays) were held in 1969. June was decided upon, despite its record of wet weather in the Waikato, because it was the month in which most farmers could take time off. The theory was that farmers would be less deterred by rain than any other social segment in the country.

The first event was modelled on the Australian National Field Days held in Orange, New South Wales, but its aims were spelt out ('broadly') as follows:

1. To increase public awareness of agricultural research, production, machinery and marketing.
2. To weld the city and country together in a major project for the benefit of all in the region.
3. To provide significant exhibitions, events and activities to attract visitors from overseas and other parts of New Zealand.

corn ($50 million) and 'mixed vegetables' ($49 million). Cut flowers, mostly to Japan, pulled in $39 million, seeds plants and bulbs $67 million, and if you wanted to count wine as a processed product of horticulture then it earned $302 million, and rising. Vegetables accessed markets throughout Australia and the Pacific islands, South-East Asia, Korea and Japan.

The domestic market is worth many millions more to New Zealand vegetable growers, though, and supply and distribution changed greatly towards the end of the twentieth century. Along with every other sector of primary industry, vegetable producers had to get bigger to survive. Fewer, larger growers were forced by the market to provide greater volumes of produce to fewer retail outlets, and directly, not through an intermediary. This caused the demise of the auction system in the early 1990s as supermarkets – reduced to just two chains in the late 1990s – consolidated their grip on the market.

The auction system, which a lot of the traditional growers valued, was too haphazard for the supermarkets. They wanted certainty and they wanted volumes. They wanted palletloads rather than boxes and crates, and they wanted supplies on 363 days a year; so the supply chain had to change and it changed dramatically. It didn't happen overnight. One of the auction floors was in action in Auckland

Most would give the Fieldays a very high pass on every count. The event developed into a commercialised but successful successor to the Ruakura Farmers' Week, which by the end of the 1960s was a famous attraction for New Zealand farmers and for some from other countries, especially Australia. Ruakura's research station was born in the nineteenth century and was reincarnated by Dr C. P. McMeekan from 1944. In 1969 it was the biggest and best-known research station in the country and one which drew 8000 farmers to inspect research projects and listen to talks by the scientists involved. Programmes most famously investigated animal management and nutrition, reproductive physiology, animal diseases, and farm machinery development. From the 1950s, Ruakura's work was noticed worldwide.

Early Fieldays were held on the Te Rapa course of the Waikato Racing Club, but within a few years the site was inadequate and the organisers bought a 160-hectare site at Mystery Creek. Numbers attending grew steadily over the years: 78,000 by 1978 and past 100,000 in the 1990s. Public accommodation for the three days in June was booked for miles around.

Sales of machinery and other, diverse, products remained fairly static through the 1990s but doubled suddenly between 2000 and 2001 to a new plateau – partly because of the breadth and diversity of the expanding programme, and partly because of a surge in farmer earnings. By 2005 the event was central to rural life in New Zealand and its appeal to city-dwellers was reflected in attendance figures as it became more folksy and fun. In addition, the Mystery Creek Events Centre was steadily developed into a superb venue.

The National Agricultural Fieldays held each June at Mystery Creek, near Hamilton, is claimed by its organisers to be the biggest annual event in New Zealand and one of the biggest for farmers anywhere.

As vegetable producers increased their scale of operations, New Zealand onions (above) became a major export to Europe, and potatoes (below) were among a number of other significant exports.

Above: The traditional auction system (pictured here in Wellington in 1949) suited small vegetable growers but its demise was brought about by supermarkets which wanted certainty and volume and organised their own contracts with the big growers.

Left: A member of the Su family harvests the lettuce crop on a farm near Levin.

for years after others closed, but that became an 'on-consignment' floor.

The effect on the growers was best demonstrated by the fact that in the early 1990s the growers' federation had more than 5000 grower members. During the following decade the number of members shrank by half. A lot of the early suppliers were small and many of them part-timers. They would deliver their produce in a small van, hand it over to auctioneers and leave it with them to sell. When the auction system disappeared the small growers didn't have a customer any more and disappeared quite quickly in the 1990s.

What happened next was that committed growers became more committed, more commercial, bigger, and the area in production went up. Yields were increasing at the same time because of the spread of the cell transplant technique from the late 1980s. When growers planted seed, they would suffer a thirty or forty per cent crop loss, but with cell transplants in the same area of ground they got a ninety per cent return. The result was crop yield boosts of around thirty per cent.

Global competition meant the pressure was always on growers. Markets for vegetable exports remained fragile in 2005 as buyers cast around the world for suppliers who would pitch their prices doßwn to clinch a sale, prices which were always powerfully affected by seasonal weather conditions in growers' countries. But New Zealanders remained dedicated to spreading the risk across a wide spectrum of produce, with entrepreneurial moves into new varieties, an explosion of growth in greenhouse salad vegetable production, and by adapting produce established in other countries to local conditions. One Canterbury grower bred purple asparagus and export demand snapped up almost all his crop. Others moved into olives, producing tautologically tagged 'extra virgin' oil of top quality; and yet others experimented successfully with truffles.

EPILOGUE

THE FUTURE: A GUESSING GAME

THE ROAD to the future is paved with the debris of confident predictions. Futurology sounded like science for a few years in the 1970s, until time ran its prophecies down in a world whose only constant is change. But agricultural trends and issues were clear enough at the beginning of the new century.

Access to international markets remained the political issue pressing most heavily on farmers, as it had been for a hundred years, although with increasing intractability since the 1960s. The nagging at developed countries to reduce barriers to world agricultural trade and subsidies to their own food and fibre producers was becoming more insistent than ever, but examination of the political stresses within those countries revealed how difficult effective progress would prove.

The mounting concern over climate change was receiving insufficient response from every country in the world because each feared putting itself at an economic disadvantage by curbing use of fossil fuels or by leaving valuable but non-renewable timber stands untouched, or selectively harvested. No one could accurately predict the ways in which climate change would affect food production or water quality and availability, although an impending escalation of water shortages threatened the peace between many countries as farming became more intensive and as population pressures meant farmers had to contest with others for their share. Border conflicts, perhaps leading to wider wars, were mooted among countries sharing major rivers and lakes that served as a common resource.

I can remember when a hose and a yard broom sent cowshed and yard effluent flowing down drains into creeks alongside which the sheds had been strategically built. The same went for effluent from sheepyards and shearing sheds. The drive for heavier carcass weights, and more intensive stocking and cropping to step up productivity, meant heavier application of artificial fertilisers, mainly nitrogenous, and that began a gradual degradation of rivers, lakes and groundwater systems in the most densely farmed regions of the country. Per-hectare application of urea fertiliser on dairy farms went up by more than 150 per cent in the closing years of the twentieth century, and by considerably more than that on intensive sheep and beef cattle farms. In the fifteen years from 1985, land under irrigation increased by close to two hundred per cent and was still climbing. Along with fertiliser application, this posed a threat to groundwater systems, many of which had become polluted. Water had been historically abundant in New Zealand but was already a contestable resource in some farming regions.

This situation prompted a 2004 call by the Parliamentary Commissioner of the Environment, Dr Morgan Williams, for a change to farming systems if traditional production was to remain sustainable. Farmers responded warily because they feared increased costs, at least in the short term. 'Intensive' is a relative word, and traditional farming in New Zealand relied more on open pasture systems than on the factory-farm systems common in North America and Europe, but the

An entrant in a competition for young handlers at the Royal Easter Show in Auckland.

intensification here was very real, with a growing reliance on heavy applications of chemical fertilisers.

The erosion of topsoil and a gradual loss of soil fertility were posing challenges. According to scientists, between 200 and 300 million tonnes of this country's soil slide into the sea every year – a serious loss to erosion on world terms. This is partly due to the crumpled nature of the landscape but was accelerated by denuding the hills of trees, by overstocking and the grazing of steep hill country. To cut soil loss, substantial areas of New Zealand land needed to be farmed under strict erosion management regimes.

Farmers were coming to terms with this issue and with the effects on the environment of the leaching of animal urine and residues of nitrogenous fertilisers into lakes and rivers. They responded responsibly. Most accepted it would add to production costs and hoped communities would share the burden. They understood that not to attend to the problem would hurt the clean green image the country promoted to sell its produce.

From the end of the 1980s the farming lobby – once impatient with any obstruction of its interests – generally accepted the fact that New Zealand life had become thoroughly urbanised and that tourism and other urban-based industries were important to national development. Farming and tourism were conjoined in the 'Farm Stay' movement. In many cases, farmers just wanted to get on with their job as efficiently as they could, rather than continually berating government and its agencies for not complying with their wishes. (Interestingly, lobbying of extreme self-interest by urban manufacturing and servicing industries had gradually taken the place once so firmly held by farming organisations.) Farming had long overlapped with other industries such as transport, the processing of food and the manufacture of packaging, and had often come to accommodate the needs of the workers involved.

A decline of once tightly knit rural communities began in the 1980s with improving roads and a fall-off in farm employment opportunities, followed by the centralisation of service facilities, such as banks, in larger towns, and the presence there of big, low-price supermarket outlets. As people left, the small towns constricted and schools and health services were also centralised. Although farmers have lost the close sense of belonging that was once so valued in rural communities, they have become less introverted in their attitudes and more sophisticated now that they have to do their business in larger towns. In addition, they better understand the views of the urban majority on such matters as the environment. Although they are concerned at the cost implications of having to control the spill-off of farm nutrients, they understand both sides of the question.

Another issue of concern was biosecurity in a globalised society with millions of people – both

The knowledge wave

In the five decades from 1930, scientific research, most of it strongly empirical, added great value to the natural climatic advantages which New Zealand's traditional agriculture enjoyed. Almost all the scientists and extension officers worked either for the Department of Scientific and Industrial Research or the Department of Agriculture. They responded in largely self-directed and innovative ways to real-life, practical problems faced by farmers, such as matching pasture varieties and fertilisers to soil types, developing animal husbandry and nutrition systems, and testing and adapting machinery that contributed so much to family-farm productivity. Teams of advisory officers were available throughout the country to disseminate scientific information.

However, from 1985 the government swept away almost all the scientific and extension support which earlier administrations had provided in various forms since the 1880s. It is fair to say that inefficiencies – even lethargy – in research and dissemination had crept into government services during the late 1970s and early 1980s, at the same time as excessive government support had disconnected the supply of products from the demand of world markets.

The government's solution was to surgically remove all supports. This drastic move was driven by a philosophical anathema to any intervention of governments between producers and markets and, consequently, the belief that farming should support its own research and development. This uncompromising attitude was underpinned by the conviction that agriculture was, anyway, a sunset industry. Thus, when Crown Research Institutes (CRIs) were set up in the aftermath of the 1980s deregulation, they were constructed as mainly trading enterprises, expected to sell their services to those who would benefit from research.

The elimination of subsidies also enabled succeeding governments to hold their hands clear and say to competing nations in which farming was heavily supported: 'See! We believe in free trade in agricultural products completely uncorrupted by subsidies.'

Some concern continued through the 1990s that too substantial an investment in agricultural research would diminish a government's ability to argue for the withdrawal of state subsidies elsewhere. However, most observers came round to the view that government research funding could surge from its 2005 level of less than one per cent of GDP to three or four per cent without the country becoming vulnerable to attack on the grounds of subsidy support.

The Labour Government of the 2000s was talking up the knowledge economy but seemed seduced by a prevailing modern myth that information technology was the end and not a means to economic development. So while it enthused about the knowledge wave, that wave seldom seemed to break on the beach, with the result that research and development remained seriously underfunded in agriculture and horticulture as much as any sector.

Scientific research in support of traditional farming in the new century was largely in the hands of the largest of the CRIs, Agricultural Research New Zealand Ltd and its subsidiaries. The AgResearch group adopted a 2020 Science campaign in 2005 with targets to be achieved over the fifteen years to 2020. It traded commercially and was making small annual profits which it aimed to lift to a 7.5 per cent after-tax return on equity. The organisation remained uncompetitive in the international scientific labour market but the ambitions of 2020 Science were bold: to double the value of dairy production and of meat and fibre production while halving both costs and the impact on the environment, mainly by developing enhanced nitrogen-fixing pastures and breeding animals for maximum pasture value and utilisation.

Visionary projects towards these ends included the development of a nitrification inhibitor in the form of a pill for cattle which could reduce nitrogen leaching by up to sixty per cent; genomic selection for animals with increased fertility and disease resistance; higher production yield from pastures with nutritional value from genomically improved grasses and clovers; the use of lipids to increase the ratio of polyunsaturated fats in meat and milk; and the enhancement of proteins in milk for a range of pharmaceutical products.

AgResearch established three new national research centres to focus on biosecurity, on animal health and reproduction, and on genomic techniques. The agency did not recoil from sensitive issues, with senior scientist Dr David Wells talking publicly of the potential to 'redesign' animals by cloning and transgenic experimentation to achieve healthier food and medical advances. Cloning produces a genetic copy of an animal and transgenic techniques produce genetically modified animals

New Zealanders and foreign tourists – and their luggage transiting sea and air ports in an uninterrupted stream. Freight also moved into the country in many forms. Add the huge length of coastline of this small nation and the menace is extreme. The protection that isolation provided for so long had given way to modern transport systems, meaning border security became more important than ever, more critical to the economy than for any other country in the world simply because of the dependence on pastoral farming. Some weeds, pests and animal diseases have already infiltrated the security screen and cost the industry dearly, but so far the major economic threats have been kept at bay. Traffic continues to grow, however, and an extension of smart border protection is essential.

Nor could New Zealand rely any longer on climate and practical farming intelligence to produce the cheapest pastoral products in the world. I recall an Australian scientist saying forty years ago that the bottom of the topsoil had not been found in some temperate-zone regions in

A technician working in Livestock Improvement's GeneMark laboratory.

through molecular biology. Dr Wells predicted that 'pharming', as it is called, could lead to elite, cloned animals for use in breeding programmes in pursuit of greater productivity. These projects were a long way from the empirical research of the twentieth century but were regarded as promising avenues for increased productivity under environmentally sustainable management.

The Horticultural and Food Research Institute (HortResearch), also one of the nine CRIs, had eleven research stations around the country and was heavily involved in biosecurity, in developing new fruit and berry varieties, and was also working in support of sustainable farming systems for fruit.

The New Zealand Institute for Crop and Food Research concentrated on sustainable land and water use for vegetables and cereals; on high-performance plants and molecular biology; on soil science (in association with other research organisations); on pest control; on biosecurity; and on the harvesting, handling and processing of seafoods.

The other six CRIs were – the Forest Research Institute, the Institute of Environmental Science, the National Institute of Water and Atmospheric Research, Landcare Research Ltd, the Institute of Geological and Nuclear Sciences, and Industrial Research Ltd. Research was supported by both Lincoln and Massey universities.

Farming and tourism are conjoined in the farmstay movement and in the setting of many top of the market lodges for tourists.

South America but that hidebound ranchers just ran stringy cattle over the land. 'If they wake up to their advantages, Australian and New Zealand farmers will have a challenge on their hands,' he reckoned. By 2005 that time had probably come.

Distance from major food markets was another impending threat to New Zealand in a world in which fuel oil was becoming increasingly rare and expensive and in which people were increasingly anxious about the effect on climate change posed by its profligate use. That was a big negative. A small positive was that if oil prices continued to ratchet up and no alternative feasible engine technology arrived in the short term, a new farming development to support the manufacture of renewable alcohol fuels looked possible – using both crops dedicated to fuel production and vegetation discarded from food harvesting and processing, such as wood chips and corn stubble.

Around the developed world, the social and financial structure of farming was also changing. Back in 1969 expatriate New Zealand economist J. B. Condliffe wrote in a book called *Economic Outlook for New Zealand*: 'In all the industrial countries there is an irreversible mechanisation of agriculture and rural depopulation. The city constituencies gain in voting power and political influence. They will not tolerate for much longer the burden of sheltering high-cost food production by small but well organised minority groups. The day of the family farm, with its supposed virtues, is drawing to a close.'

Thirty years on, most of those industrial countries Condliffe had in mind were still tolerating that high-cost burden but the depopulation of the rural areas was certainly happening, with more than 300 farms lost every week in the United States, most of them subsumed by corporations or urban developers. By early 2003 nearly two-thirds of agricultural production in the United States came from three per cent of its farms, which resulted in the extraordinary situation of more than seventy per cent of government support money going to ten per cent of farming operators – all of them among the largest producers.

Could that scenario happen here? Now, a hundred years on, would a reversal take place of the breaking up of the large estates into smaller units? For many people, both locals and foreigners, the family farm in New Zealand provided lifestyle opportunities in a relatively benign environment unequalled anywhere else in the world. But business is unsentimental, and already traditional farming enterprises in this country are becoming bigger and fewer, attracting capital from non-traditional sources. Twenty years after deregulation the pressure was on many small farmers to get bigger or get out, but mainly in dairying where opportunities to automate

Dog-trialling has become a spectator sport for tourists as well as for country aficionados, with farm families hosting both urban domestic and international visitors.

were greater. More than one dairy farmer, however, was looking at a maximum of 800 cows per farming unit as the economic optimum.

Farming by corporations in the United States was made eminently feasible by the horizon-to-horizon tracts of flat land planted in cash crops that were mechanically harvested and processed in such huge volumes that no small farmer could hope to compete. The massive domestic market for grains and meat also gave an incalculable advantage to the corporation with its economies of scale.

So the big question was not whether farming would continue to be the mainstay of New Zealand economic life but whether family ownership and management could survive in the face of outside, perhaps corporate, investment. Conditions here are vastly different from those in the United States and Europe, and some of them work in favour of a continuation of the family enterprise. Tourism had become a support industry for many farmers through the farmstay movement that began in the 1960s and gradually grew. Farm families – mostly those not involved in the diurnal demands of dairying – hosted both domestic and international visitors, and provided them with entertainment through farm activities such as shearing, handicrafts and dog-trialling, as well as experiences of hunting, fishing, jet-boating and, seasonally, skiing. The strength of these enterprises was the family, and their importance may grow in an increasingly crowded world.

While rural folk pondered what would happen, thousands of people from the cities moved into the surrounding countryside, either to live in what were called 'lifestyle blocks' or to visit them on weekends and holidays while living in apartments in nearby towns – a way of life made increasingly possible by big cars on improving roads. An argument pivoted around whether the subdivision of good-quality farming land seriously reduced production, or whether output would in many cases be improved by those on small blocks who used the land for intensive farming of free-range chickens, for horticulture or for high-value organic products. Did the future lie with thousands of boutique farms producing high-value products alongside large, corporate pastoral and cropping enterprises? These issues hung in the balance, and while many were pre-pared to guess no one was tempted to offer an adamant answer. But if the past is anything to go by, New Zealand farmers will adapt and innovate and continue to lead the way.

Acknowledgements

I talked with many people researching this book and I am grateful for their collective wisdom. I would like, however, to thank especially Clive Dalton, David Yerex, Neville Martin, Mick Calder, John Scott, Jeanine Graham, Tui Flower, Jenny Jones, the Raine and Mathis families, Ian F. Grant and Ken Robertson. Special thanks to Wendy Canning of Sydney, and Maureen and Ron Redwood of Russell for providing me with very congenial spaces in which to write; and to my wife, Dawn, for her encouragement and for giving me the mental space in which to write.

Photographic Credits

All new colour photography by Ian Baker, unless otherwise credited.

Key T – top, B – bottom, L – Left, R – Right
AFFCO 163T. **Alexander Turnbull Library** 4, 17, 18, 23, 24, 25, 27B, 29, 30, 31, 39T, 43, 46, 50, 53, 54, 55, 56, 58, 66, 70, 79, 84–5, 91, 98, 100-1, 106B, 107B, 112, 114, 116, 117B, 130–1, 132, 135L, 137, 139R, 139L, 144, 150T, 151, 165, 171, 206, 218, 234, 237TL, 239, 250, 251; Agriculture Department Collection 74-5, 122; Bennet Iorns Collection 47; C. Johnston 62; Dominion Post Collection 11, 115, 153, 155, 161, 169, 222 (courtesy Maggie Crump), 223; Evening Post Collection 94L, 131, 220, 248L; G. L. Adkin Collection 67, 110B, 233; Gordon Burt 135R, 248R; Greig Collection 44–05; H. J. Schmidt Collection 165; James McAllister Collection 117T, 121, 141, 142, 180–1, 195; J. H. Grigg Collection 97; John Pascoe 88, 152, 172; Making New Zealand Collection 20, 26, 42, 52, 60, 69, 86, 92R, 93, 113, 134, 162, 200, 230, 237B; New Zealand National Publicity Studios Collection 9, 10, 92L, 184, 232, 179, 190, 217, 232, 236, 238, 249, 253R; New Zealand Free Lance Collection 96, 214, 245; Northwood Collection 16, 105, 110T, 119, 229; Rhodes Album 48; R. P. Hargreaves Collection 108–9; Russell Orr Collection 201T (courtesy of Julia Gillie); S. Head Collection 71; Sinclair Collection 103R; S. P. Andrew Collection 63B, 106T; Steffano Webb Collection 31, 150B, 201B; Tesla Studios Collection 27T, 128; The Press (Christchurch) Collection 75; W. H. Raine Collection 138. **Arno Gastieger** 81. **Corbis** 148–9, 225, 246, 253L. **Deer Industry NZ** 224, 225R. **Hawke's Bay Museum** 85. **Hocken Library** 28, 39B, 57L. **Livestock Improvement** 129, 257. **Massey University** 102, 203. **Museum of New Zealand Te Papa Tongarewa** 16, 22. **Otago Settlers Museum** 64, 94L. **Phillip Holden** 2–3 133, 209 (courtesy of Leon Russell-White). **PhotoLibrary** 214. **Photo New Zealand** 252. **Puke Ariki** 124. **Rex** 146–7. **Rex Forrester** 222 (courtesy of Roger Forrester). **TVNZ** 243. **Wairarapa Archive, Masterton** 49.

Index

Note: Italicised page numbers indicate photos.

abattoirs 93
Aberdeen-Angus *see* Angus
aerial seed-sowing 114, 115
aerial topdressing 114, 115; DC3 *115*
afforestation 230
AgResearch 256
Agricultural Mission School 23
agricultural, expertise 24; revolution 38, 52; techniques 23
agriculture 21, 22; market research 168; Maori 23, 24; mixed 38
Alley, Rewi *135*
alpacas *159*
Anchor 123
Anderson, John 66, 69
Angora goats *202*; fleece 225
Angus 214, *215*, 215
animals, draught 21; introduced 87; pastoral 21, 24, 32
Anthony, Frank S. 134, *135*, 135
Apple and Pear Board 237
Apple and Pear Marketing Board 239
Apple and Pear Restructuring Act 240
apples 234, 235, *238*; international markets 238; orchard 239; varieties 237
apricots 235
aquaculture 159
Auckland 77
avocados 244; export 244
Ayrshires 189, 192, *198,* 198

Banks, Joseph 19
barley 23, 27, 76
beef, export quotas 219
berries 16
berryfruit 247
biosecurity 86, 234, 256, 257
birds 21; land 16, 17, 19, 20; sea 16; trapping 19
blackcurrants 247
Bledisloe, Lord *138*, 138, 142
blueberries 247
Border Leicester *98*, 99
Bowen, Godfrey *206*, 206; Ivan *206*, 206
boysenberries *247*
branding 52, 161, 162, 164; apples and pears 239; butter 123, 124; national 123; venison 224; Zespri 244
breeding, animal 98; artificial *129*, 129, 197; cattle 98; revolution 99; sheep 97, 98
breweries 65
Brown Swiss 192
bulk, buying 123; exporting 154, 160; purchasing 151, 166
bullocks 57
burning 16, 105, *108–9*, 228, 230
burn-off 17, 111, 112
Busby, James 244, 245
bush, country 107; farmers 112, men 106; sickness 141, 230
Butler, John 26
Butler, Samuel 52, *53*, 53
butter 116; exports 77, 118, 121, 123; factories 118, 124
butter-making 122, 128, *131*; experts 121

cabbage 172
cabbage tree *14*, 14
calicivirus 87
canoes 28; transport 25
Canterbury 46, 47, 48; grain 65; Pilgrims 50, 53; province 60
Canterbury Association, The 52
Canterbury Frozen Meat Company 96, 97
Captain Cookers 88
Cargill, William *64*, 64
carpet industry 102; manufacturing 207
Carter, Charles Rooking 47, 49
Carterton 49
cashgora 225, 226
cashmere 225, 226
cassava 18
cattle 22, 57; breeds 215; exotic 214; numbers 90
cell transplant 252, 253
cervena 223, *224*
Charolais 215
cheese 116; exports 77, 118, 121, 123
cheese-making *122*, 122, 129
Cheviot Hills Station 100, 101
children 108, 112, 113, 134
chilled beef 217, 218; marketing 218
chilled meat 206, 210
Chinese gooseberry 242
Chong, Chew 123, *124*, 124
Church of England 46, 52, 97, 116
climate 32; change 254, 258; continental 32
cloning 256
Closer Economic Relations 153
clover 23
Cockayne, Dr Leonard *142*, 142
cold storage 164
colonists 46, 61
conservation 105
Cook, James 17, 88
Coopworth 101, 202
Correspondence School 113, *114*
Corriedale 99, 101
Country Calendar 243
creameries 123, 125
crops, cereal 30
Crump, Barry 221, 222
cultivation 17; machinery 65

Dairy Board 183; legislation 185
Dairy Goat Co-operative (NZ) Ltd 226
dairy industry 31, 116, 118, 123, 183, 196; marketing 123, 129
Dairy Research Institute 143
dairy, co-operatives 119, 122, 123, 124, 125, 143, 184, 189, 196; exports 110, 142; factories 118, *121*, 122, 125, 128, *185*, 185, 196; imports 118
dairying 118; Banks Peninsula 118, Southland 121; Taranaki 116, 121
Darwin, Charles 23
Davidson, William Saltau *94*, 94
Deans, John 42, 55
Deans, William 42, 55
deer 87, 158, 220, 221; breeds 220; farming 221, 223, 224; red 220, 223; velvet 224, *225*
deforestation 16, 105, 228, 230
Depression, the 150, 154, 165, 178, 230
depression 48; agricultural 29; economic 67, 90, 143
deregulation 162, 163; domestic 158
Dieffenbach, Ernst 42
digging stick (ko) *16, 18*, 30
diseases 60, 86, 234, 257; catarrhal 60; contagious 212; control 237; exotic 214; fungal 234
dogs 16, 86, 88; wild 86
Drysdale 102, 202
Duggan, Eileen *106*, 106, 133
Dunedin 94, 96
Durhams 116, 214

economy 10, 11, 153; domestic 165; deregulation 132, 155, 170, 185; knowledge 256; New Zealand 90, 94, 158
Edendale 119; Estate 121
education 113; agricultural 140, 147
eels, catching 19
electric fences 131, 132
elk *see* wapiti
emigrants 46, 47, 90
Enclosure Acts 137
Endeavour 17, 19
English Leicester 99
environmental, dangers 86; pollution 195, 197
Enza 239, 240
Enza Ltd 240
erosion 113, 255
European settlers 17, 32
extractive industries 77, 228, 231

facial eczema 147
Fagan, David *207*
families 112, 113, 133, 134, 182
Faraday, Michael 91
farming 19, 21; arable 39, 102; commercial 21, 140; hill-country 114, 115; intensive 110, 115; lobby 61, 158, 255; Maori 28; mixed 73; pastoral 17, 36, 39, 70; pre-European 14; regions 33–5; subsistence 21, 39, 47, 49
farmstay 258, 259
feathers 21
Featherston 49
Federation of Taranaki Co-operative Dairy Factories 143
feijoas 244
fern root 16, 18, 21
Fernleaf 123
fertilisers 102, 115, 141; artificial 254; chemical 255
Finn 205, 208
fireblight 238
fish 16, 20; drying 19; hooking 19; netting 19; storage 19
fishing 19
flax 16, *20*, 23, 231; baskets 20; blankets 20; clothing 20, 21, 231; cordage 231; export 231; seine net 20
flowers, cut 250
Fonterra 185, 189
foot and mouth disease 212

forests, beech 32; exotic 34, *36*, 230; indigenous 228, 231; North Island 104, 105; plantations 232; products 151
fowl 16, 88
fowling 19
Fraser, Miss Isabel 242
Free Kirk Scots 50
freezing, industry 162; workers 9; works 93, 96, 102, 162, *163*;
Friesians 191, *199*, 215
frozen meat exports 162; trade 163
fruit 21; bottling *249*; canning 248, 249; freezing 248, 249; industry 247, 248
fruits, citrus 235, *236*; pip 235; stone 234, 235, 248; temperate-zone 234
fungicide 115

gatherers 16
Gear Meat Company 218
genes, new 214, 215; pool 215
genetic material 208, 226
goats 87, 224, 225; milking 226; milk marketing 226
gold 57; export 64; miners 80
goldrush 64, 65; Central Otago 64; Westland 64
gorse 78, 79, *80*, 81
gourd 14, 17, 18
grain 18, 27, 65; boom 64; exports 47, 65, 71; Maori-grown 23, 24; Maori-milled 23, 24
grape-growing 244, 245
grapes 158, 234, 240, *245*; varieties 247; vines 235, 246
grass 70, 129, 132; seed 112; sowing *112*
greenstone 21
Grey, Sir George 47, 49
Greytown 47, 49
gridironing 61, 97
Grigg, John 96, *97*, 97, 192
Group Breeding Scheme 203
Guernsey 192
gumdiggers 105, 229
Guthrie-Smith, Herbert 83, *83*, 87

harrow 71
harvesters, header- 30; McCormick 66
harvesting 19, 20; machinery 65
Hawke's Bay 46, 47, 48
hay 129, 132, 133; baler 70
haymaking 30
Heaphy, Charles 42

helicopters 115, 220, *222*, 222
Hereford 214, 215
Highland cattle 213
Hika, Hongi 23, 25, 26
hill country 113, farming 114, 115
Holstein-Friesians 189, 191, 192, 195
Holyoake, Keith 160, *161*
hops 35
horses 22, 86; numbers 90
HortResearch 257
house cows *114*, 116, 117
hunters 16, 220, 221, 222, 225
hydatids 87

immigration 90
implements 30; iron 21
Industrial Revolution 38, 52, 137
infrastructure 90, 94
insecticide 115
internal combustion engine 131
irrigation 197, 254; spray 132
itinerant workers 45, 58

Jerseys *189*, 189, 191, 195, *199*

kaheru 30
kahikatea 228
Kaingaroa Forest 230
karaka 16
karakia 19
kauri 104, 105, 228; sap 229
kauri gum 77, 228, *229*; exports 228
Kawerau 231, *232*
Kirton, Dr A. H. 166, *168*, 205
Kiwi (group) 185
kiwifruit 158, 234, *241*, 242, 244
ko *see* digging stick
kumara 14, *15*, 17, 18, 19, 20, 21; gardens 17; harvest 18; Maori 14; planting 18; red 14; soil preparation 18; storage 18, 22
kuri 86, 88

Labour Government 73, 123, 139, 150, 155, 256
labour, seasonal 58
Land Girls 150, *152*
land, first sale 26; ploughing of 70; policy 138; sustainable use 257
landscape, North Island 36; South Island 6
Lange, David *155*, 155

leasehold movement 139
Lee, John A. *63*, 63
Leicester 98
lifestyle blocks 259
Lincoln 99, 111; lambs 112
Lincoln School of Agriculture 140, 147
Little, James 99, 101
livestock 16; cargo 24
Livestock Improvement 192
logging 105
Lyttelton 50

machines 30; farm 131; milking *31*, 31, 190, 195; shearing *30*, 30, 31; threshing 66
McKenzie, John 138–9
MacLoughlin, Jim *240*, 240, 242
McMeekan, Dr C.P. (Percy) 129, 131, 143, *145*, 145, 251
macrocarpa 230
mad cow disease 212
maize 18, 24, 76, 137
Manpower 150
Maori 14, 17, 21; exodus 151
marketing 164; bulk 168; campaigns 150
Marlborough 46, 47, 48
Marsden, Samuel 21, 22, 23, 24, *116*, 116, 244
Marshall, Jack 166, *169*
Massey Agricultural College 102, 143
Massey, William *164*, 164
Masters, Joseph 47, 49, 49
Masterton 47, 49
meat, canning 91, *92*; export 99, 110, 166; preserving 91, 92; price 103; tinning 91
mechanisation, farm 47
merinos 39, *40–1*, 53, 57, 96, 98, 99, *103*, 202, 205; fleeces 55
Mesopotamia *52*, 52, 53, 54
Middle Island 50
military force 31
milk 116, 128
milking 116, 117, 133; machines 117; once a day 197
milking shed, herringbone 190, 195; rotary system 190, 195; turntable 190; walk-through 190
mill, flour 27, 69
millet 18
missionaries 21, 22, 24, 26

moa 16, 17
mohair 225
Monrad, Bishop Ditlev *106*
mowers, mechanical 30
Muldoon, Robert *153*, 153
mussel crop 159
myxomatosis 87

National Agricultural Fieldays 250–1
National Development Conference 160, 161
National Flock Recording Scheme 202–3
National Government 151, 153
New Zealand and Australia Land Company 94, 99, 119
New Zealand Company, the 38, 39, 42, 49, 52
New Zealand Co-operative Dairy Company 123, 143
New Zealand Dairy Group 185
New Zealand Dairy Produce Export Control Board 123
New Zealand DSIR 142
New Zealand Meat Preserving Company 92
New Zealand, cuisine 173, 176; infrastructure 150; population 57, 64, 90
New Zealand-Australia Free Trade Agreement 153

oats 18, 27, 73, *74–5*, 76; yields 69
obsidian 21
olive plantations *177*
Opouriao Dairy Company 170
Orchard and Garden Diseases Act 234
orchards 27, 29, 234, 237
ostriches *156–7*, 158, 159
Otago 47, 48; grain 65; population 64
oyster production 159

palm fronds 16
Pani 18, 19
Papa the Earth Mother 19
pasture, types of 141
peaches 234, 235; trees 27
pears 234, 235, *237*, 237; varieties 237
Peninsula Cheesemaking Company 119
Perendale 101, 202
persimmons 244; export 244
pests 87, 89, 257; control 86, 237, 257
phosphates 132

phylloxera 246
pigs 16, 21, 88; exports 88; farming *88*, 88
Pioneer Butter Company 119
planting 19, 20
plough 26; double-tine 26; multi-tine 72, 73; single-tine 110, two-tined 70
Polynesian settlers 14
possums 87, *89*, 89; skins 89
potash 70, 111
potato 14, 17, 18, 20, 21, 23, 27, 178, 179
Poverty Bay 48
power, combustion systems 31; electric 31; steam engine 31
Preston's Famous Cow Detectors *128*
primary products 160, 165
public works 90
pulp and paper 231

quality 128; control 123
quarantine 86; laws 60

Rabbit Nuisance Act 80
rabbits 47, 58, 78–85, 87; decommercialisation of 85, 87; exports 82, 85, 87; farming 82; skins 82, *84*, 85, 87
radiata pine 230, *233*
Rae, A. L. 202, *203*
railways 47, 105
Raine, Richard 216
rainforest 32, 104
Rangi the Sky Father 19
rats 16, 88
reafforestation 17, 113, 228, 231
reapers 66; and binders 66, *67*, 67; Burgess and Key 66; McCormick 66
reaping 66; machine 65
refrigerated shipping 90, 91, 92, 96, 102, 110
refrigeration 31, 93, 103
remittance men 63
Research and Development 256
revolution, food 176; social 170
Reynolds, Harry 123
rice 18
rimu 228
roads 105
Robinson, William 62
rock guano 141
Romney *96*, 99, 202; fertility 101, 102; fleece 102

Rongo 18, 19
Roslyn Worsted and Woollen Mills *150*
Ruakura Agricultural Research Station 251
Ruakura Animal Research Station 131, 143, 145, 166
Ruatara 22, 23, 25
rural communities 256

salmon crop 159
Sawers, John 122
Saxton, John 216
scab 43, 47, 58
schools 182
schoolteachers 113
scientific research 140, 142
Scottish blackface 102
scrapie 208, 214
scythes 30, *65*
sealers 21
seals 16, 17, 20
seed, treatment 73
seine net 20
settlers *104*, *107*
shagroons 53, 55
share-milking 124
shearing gang *201*
shears, blade 30
sheep 22, 57; farming 39, 47, 66; imports 51; meat 47; numbers 90; shearing 50, 51, *59*, *204*, 205–6, *206*; varieties 208
sheep industry, marketing 155
Sheeplan 101
sheepmen, Australian 52, 55
shellfish 16, 21
shorthorn 191
sickles 30
silage 30, 129, 132, 133
Simmentals 192, *213*
skimming stations 123, 125
social change 21
soil 32, 33, 140; analysis 141; deterioration 140; North Island 35; preparation 18; science 257
Southdown 98, 202
spotting 61
squatters 42, 43, 46, 52, 55, 60, 140; Australian 55
squatting 52, 53, 57, 60
steam traction engine 70
steamships 94
stockmen, Aborigine 55

Stratford District High School *141*
stud breeders 101, 202, 203
subsidies, farm 154
supermarkets 251, 253
superphosphate 70, 73, 103, 115
supplementary minimum prices 153, 154, 155, 163
Surville, Jean Francois Marie de 25, 88
swaggers *63*, 63
sweet potatoes 18

tamarillos 242, 244
Tane 19
Tangaroa 19
tapu 21
Taranaki War 24
Taranaki wool 124
taro 14, 17, 18
Tasman Pulp and Paper Co 230, *232*
Teviot Station 55, *56*
Texel 205, 208
Texel/Perendale *205*
textile industry 226
Tiki 19
timber 16, 39, 77, 228; imports 228; industry *105*, 105; mills 232, 233
tohunga 18
tools, metal 22
topdressing 132
tourism 11, 158, 163, 255, 258, 259
traction engine 86, 201
tractors 26, 30, 73
trade 21, 28; export 23, 38; fat lamb 73
trading 19
transmissible spongiform encephalopathy disease 208
trapping 19
Treaty of Waitangi 49
tree tomatoes *see* tamarillos
tree-felling 17, 70
tuberculosis 89
turnips 21, *180–1*

Utopia *144*, 144

vegetable growing 249; industry 248, 250
vegetables 21; domestic market 251; fresh exports 250, 253; processed 250
venison 224; export 223; marketing 223
vineyards 246

Wairarapa 42, 46, 47, 48

Wairarapa Small Farms Association 47, 49
wapiti 223
water quality 197
watercress 18
Wattie, James *248*, 248
Watties 248, 249, 250
weapons 16
weather 21
Weld, Frederick *42*, 42, 43, 46
whalers 21
Wharekaka Station 43, 43
wheat 18, 21, 23, 24, 26, 27, 29, *68*, 69; boom 23, 66; exports 71; prices 67, 69; varieties 73; yields 70, 71
Williams, William 24
wine 170, 173; export volumes 247; industry 247
wine-making 245, 247
wives 9, 112, 113
women 47, 48, 58, 65, 114, 134, 153, 174; ; Maori 20
Women's Division of Federated Farmers 134
Women's Institute *134*, 134
Women's Land Service 150, 152
wool 39, 47, 200; buyers *200*, 200, 201; exports 47, 90; local market 64; marketing 207; merino 38; press 49; prices 48, 52, 64, 65, 67, 90, 103, 153; trade 102
wool-growing 38
Woolmark 200
workers, itinerant *44–5*, 58, 62, 63
World War Two 123, 150, 153
Wright, Hayward 242

yams 14, 17, 18

Zespri gold 244
Zespri Group Ltd 244